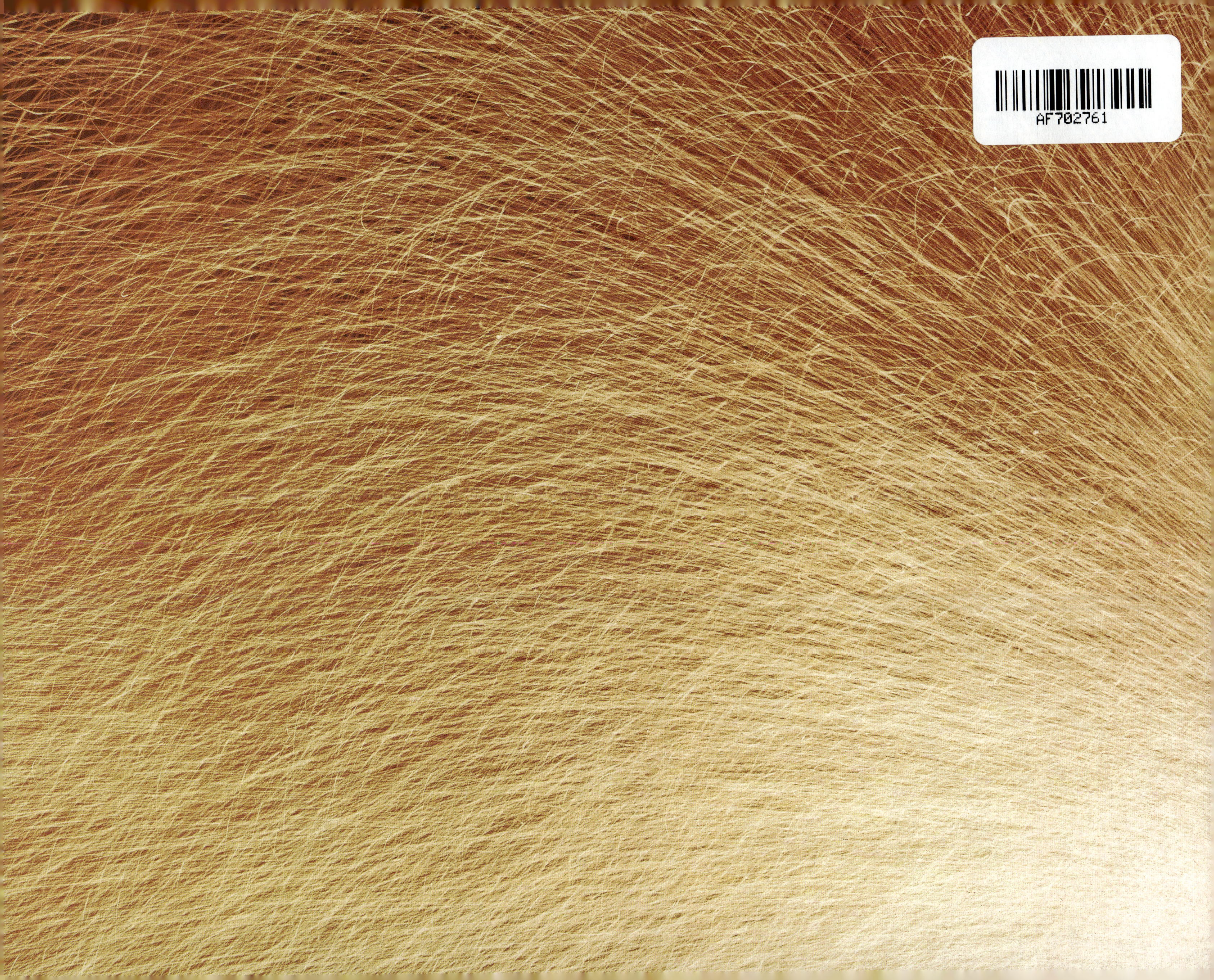

Inside BETHLEHEM STEEL

PT Photo Books
917 Highland Avenue
Bethlehem, Pennsylvania 18018
U.S.A.

www.ptphoto.com

ISBN 978-0-9798657-0-1
LCCN 2007936653

1. Photography.
2. Art.
3. Industrial History.
I. Treiber, Peter B.
II. Kovach, Elizabeth A.
III. Title.
IV. Bethlehem Steel.

This book was designed by
Kathleen Forsythe, Forsythe Design Inc.
www.forsythedesign.com
Edited by Jenifer Gamber
Typeset in Rockwell and Nobel
by Hope Xinyuan Book Printing Company
PT Photo Books can be contacted through
www.ptphoto.com

Inside BETHLEHEM STEEL

The Final Quarter Century

Photographs by **Peter B. Treiber**

Text by **Elizabeth A. Kovach**

With Introduction by Elizabeth A. Kovach

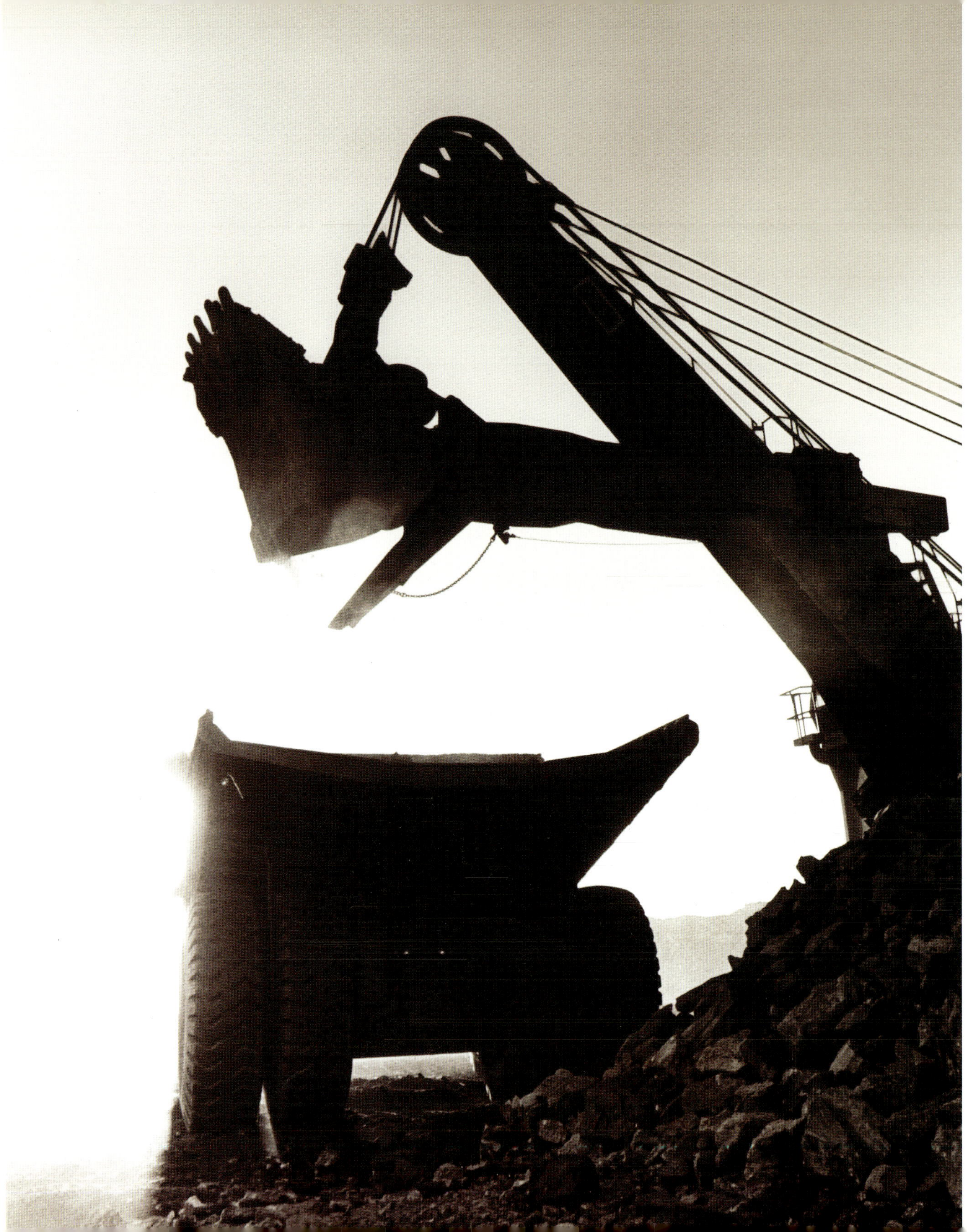

2

Loading taconite in Hibbing

Sentinels of Safety

Bethlehem Steel emphasized safety throughout its operations. These miners stand before the Sentinels of Safety Award, the most prestigious safety award in the mining industry. Presented by the National Mining Association annually since 1925 to the safest mine in the United States, the award recognizes outstanding safety records, stimulates greater interest in safety, and encourages development of more effective accident prevention programs. Bethlehem's mines were awarded the Sentinels of Safety on numerous occasions in the 1970s and 1980s, a testimony to the value Bethlehem placed on safety awareness and training, employee diligence, and proper, well-maintained safety equipment.

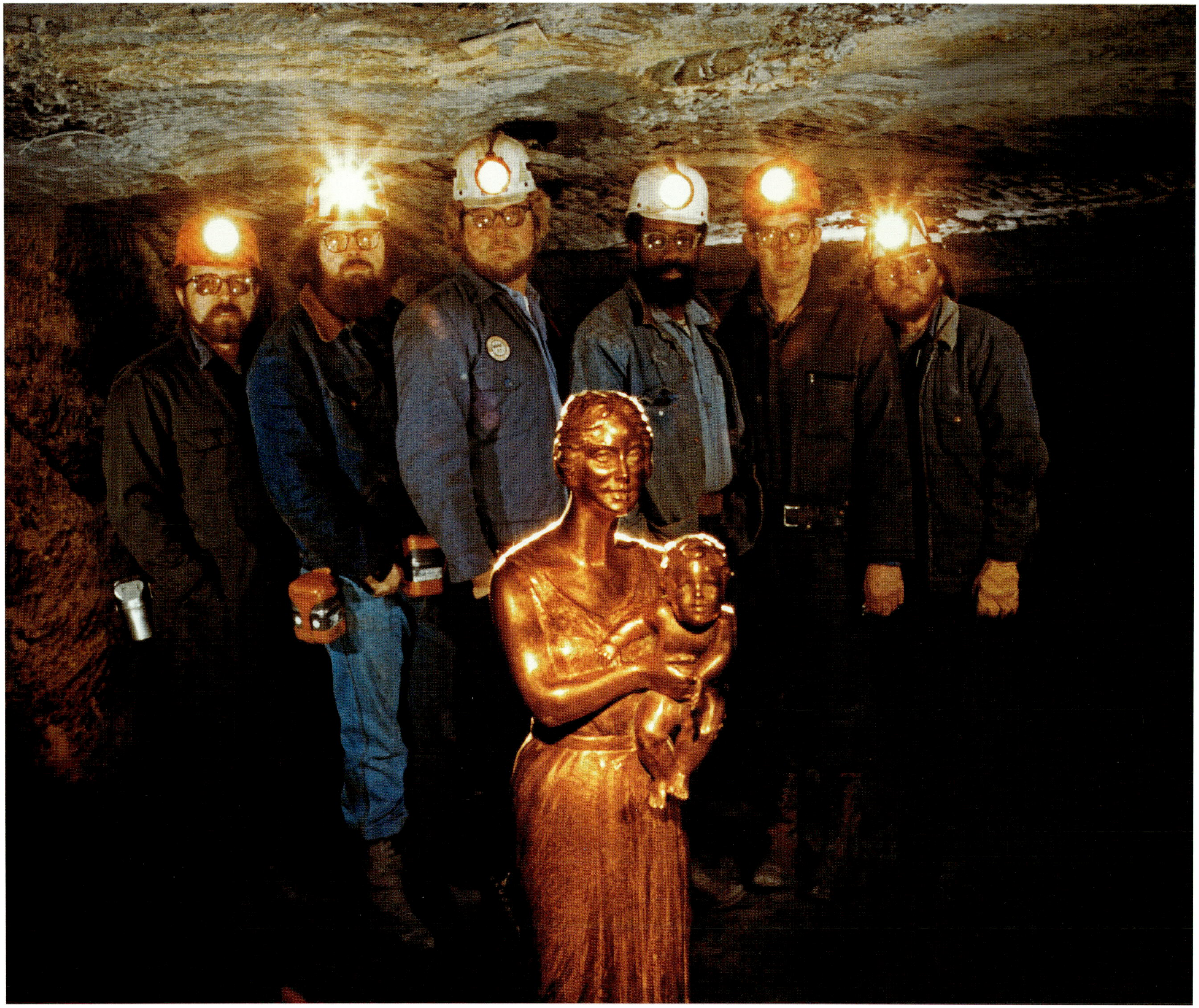

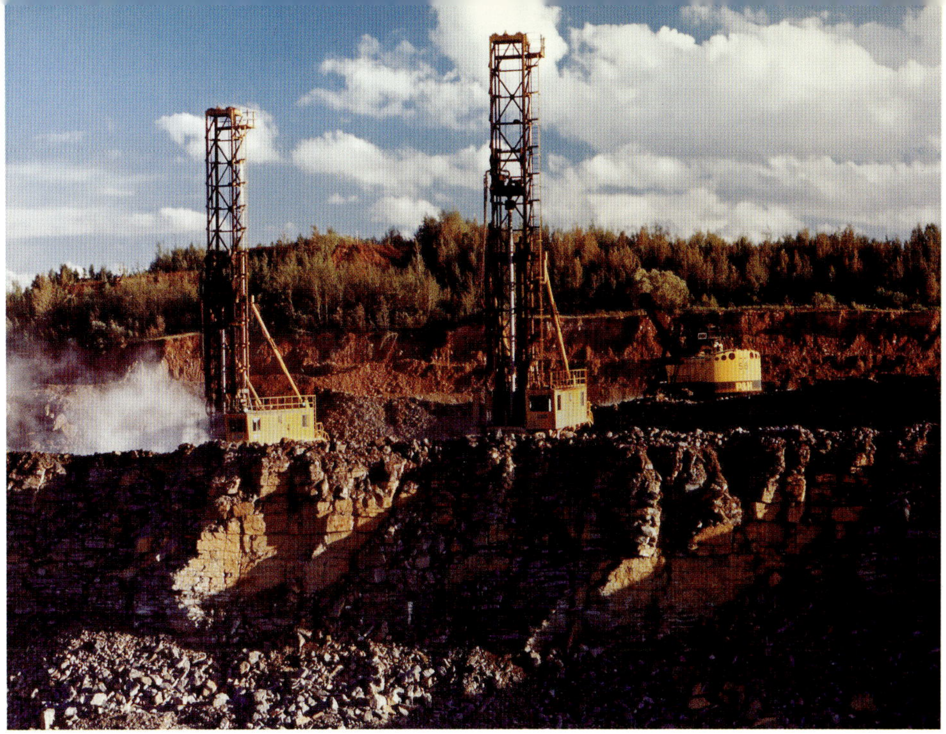

Drilling blast holes,
Hibbing, Minn.

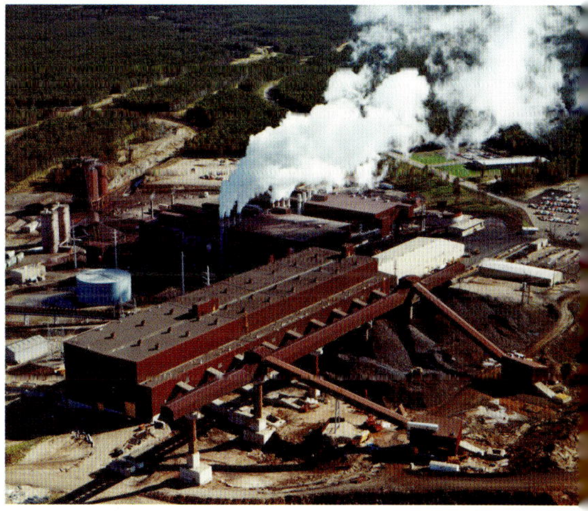

Hibbing
pelletizing plant

Large drill rigs bore 16-inch diameter holes up to 40 feet deep in a blast pattern to expose the taconite at Hibbing, Minn. The holes are then filled with blasting agents. Once all workers and equipment are cleared, the holes are detonated a millisecond apart to create an avalanche of ore. Hydraulic shovels load the ore into trucks that weigh more than 200 tons. The ore is then taken to the preparation plant to separate the debris from the mineral-rich material, which is formed into iron ore pellets.

Bench blast

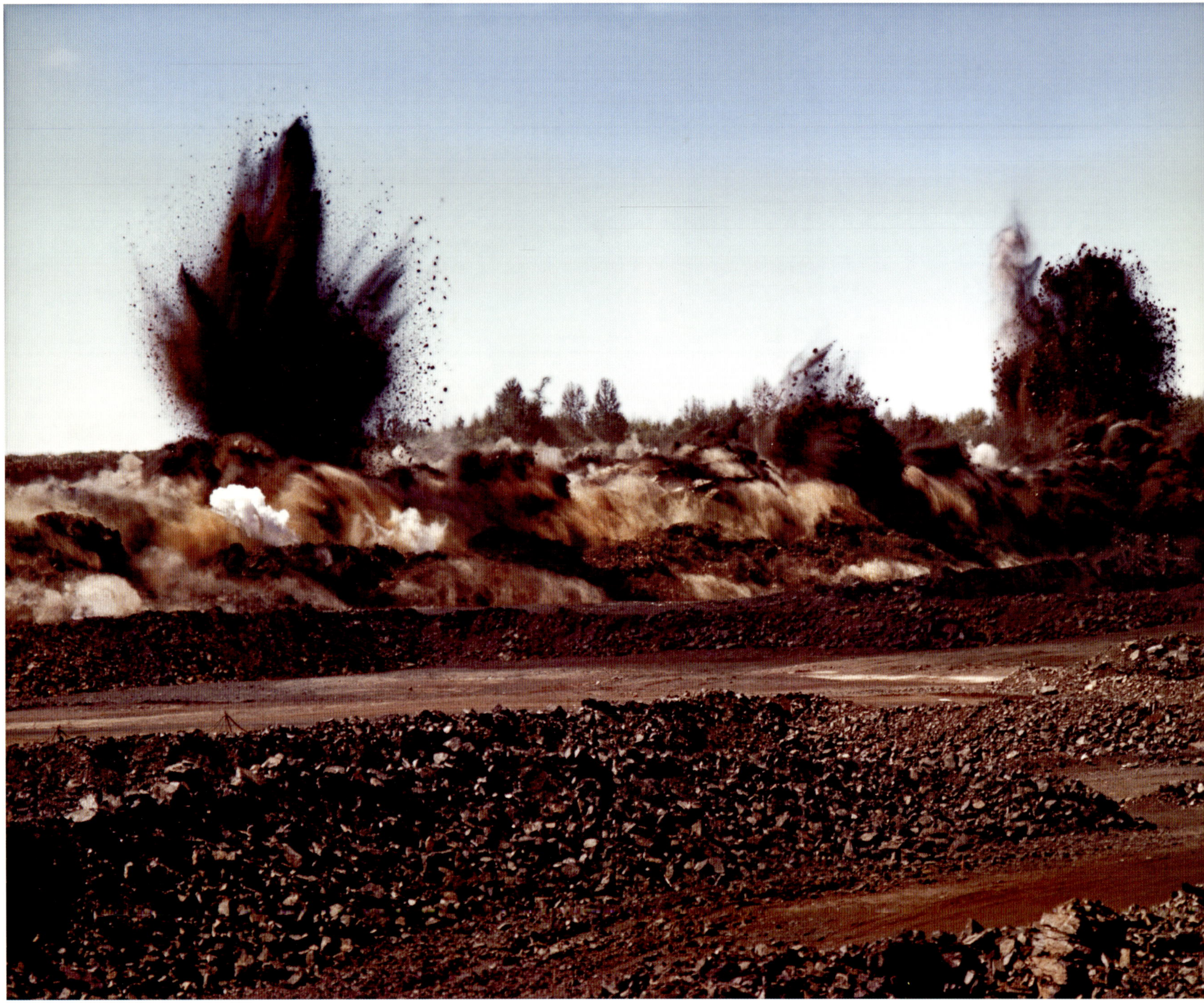

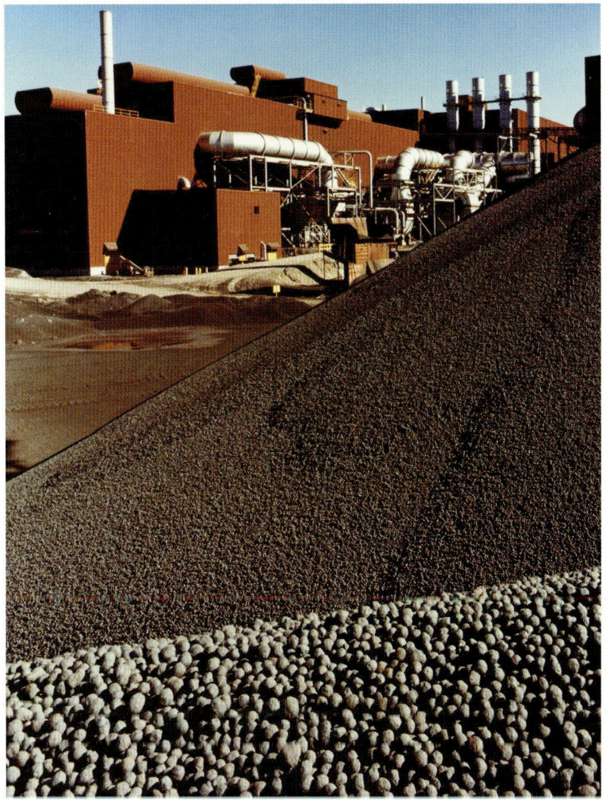

1

Hibbing iron ore pellets

1 Cleveland Cliffs' Hibbing taconite operation, located in Minnesota's Mesabi iron range, has a pelletizing plant that takes lumps of taconite, an iron ore, and forms little ball-shaped pellets about 3/8 inch in diameter. The pelletizing process saves the mineral-rich ore and discards unwanted stone that would not melt when injected into the bell of a blast furnace. Before Hibbing began producing taconite pellets in 1976, Bethlehem received iron ore from other Cleveland Cliffs' mines.

2 At the Hibbing, Minn., preparation plant, large gyrator crushers reduce the taconite to chunks six inches or smaller. The chunks are further reduced to 3/4 inch or less by ball mills with four-inch steel balls, such as these seen here. Once appropriately formed and sized, the pellets are charged into a rotary kiln to stabilize their shape to endure the transport and delivery to waiting blast furnaces.

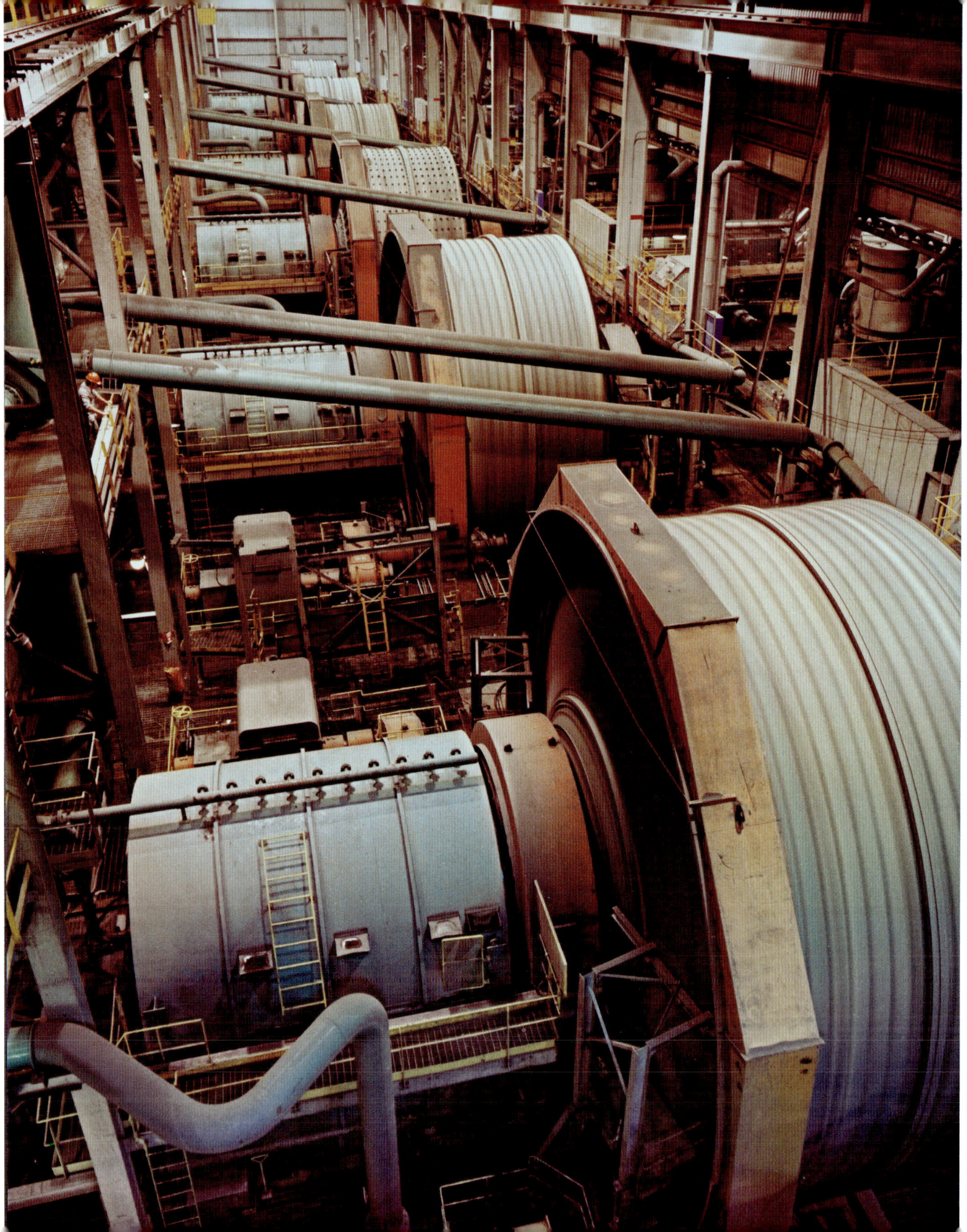

2

Hibbing ball mills

Offloading coal at Burns Harbor

Raw materials are off loaded at Burns Harbor, Ind. from vessels that traverse the Great Lakes from the upper midwest and Canada. The fully integrated steelmaking facility at Burns Harbor uses coke, iron ore, and limestone in its blast furnaces to create iron, which is later refined into steel. Burns Harbor uses iron ore pellets produced by Cleveland Cliffs' taconite mines and pelletizing plant in Hibbing, Minn. The pellets are shipped by rail to Duluth, Minn., and then loaded onto ore vessels, which, after being fully loaded, travel about one full week to the Burns Harbor plant at the southern tip of Lake Michigan in northwest Indiana. Bethlehem's fleet, called the Great Lakes Steamship Company, made the seven-day route from Minnesota to Burns Harbor. Two vessels in the fleet were the *MV Stuart Cort* and the *MV Lewis Foy*, both named after retired chairmen of Bethlehem Steel. Bethlehem Steel had a tradition of naming important corporate assets after retired chairmen. The research center in Bethlehem was named after Arthur Homer, and the last headquarters building was named after Edmund Martin.

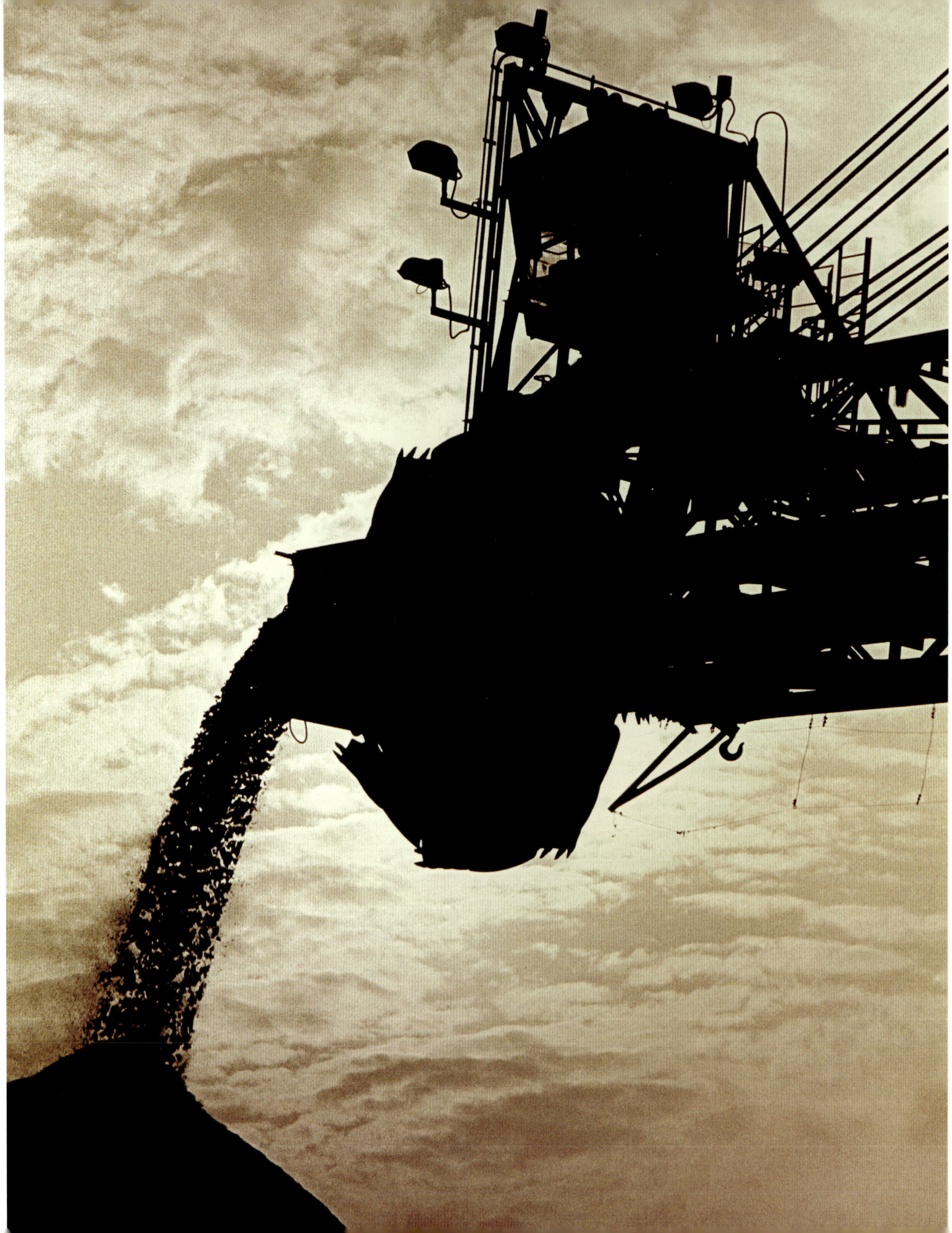

2

Coke byproduct storage

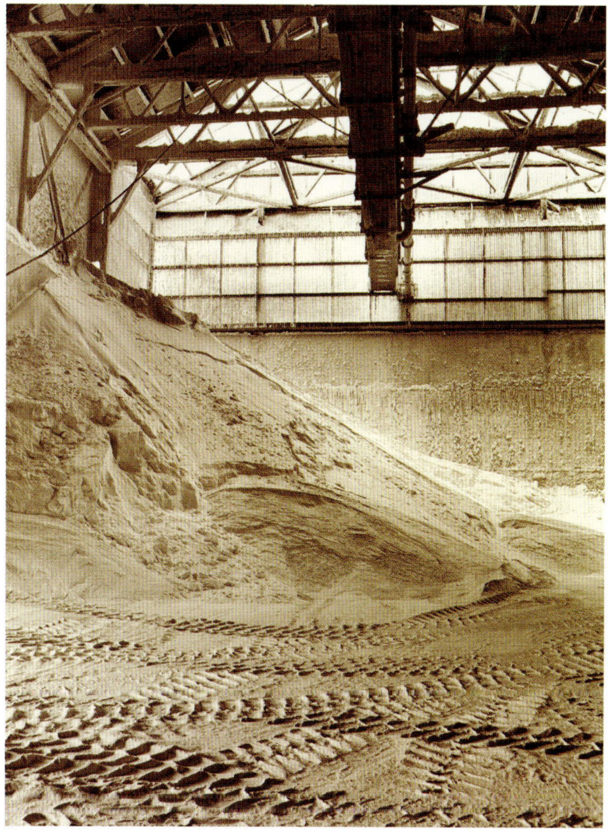

1 Cokemaking is the first step of the fully integrated steelmaking process. Coke is formed from soft or bituminous coal that is baked in vertical chambers called ovens, which are grouped into batteries of as many as 50 ovens. Each oven is lined with refractory bricks. Once the coal has been converted into coke, it is pushed from the oven into a covered rail car and then cooled, or quenched, with water. For many years, Bethlehem Steel owned vast coal mining operations in Pennsylvania, West Virginia, and Kentucky. The subsidiary, BethEnergy Mines Inc., provided the vast majority of coal for coking at Bethlehem's fully integrated plants.

2 Coking is an essential part of the steelmaking process, but, as those living near the ovens surely know, it is not without its byproducts. Some of the byproducts created pungent smells and airborne soot, and, beginning in the early 1970s with the initiation of the Clean Air Act and the founding of the Environmental Protection Agency, steelmakers were told to "clean up your act," literally and figuratively. Bethlehem, like all other steelmakers, spent hundreds of millions of dollars to install technology to contain air and water emissions. But even before the mantra of "recycle and reuse" was fashionable, the steel industry had learned it could capture and sell coke oven byproducts, as seen here, to manufacturers of products such as fertilizers, plastics, dyes, synthetic fabrics, and drugs. Even perfumes contain byproducts of the coking process.

1

Pushing coke

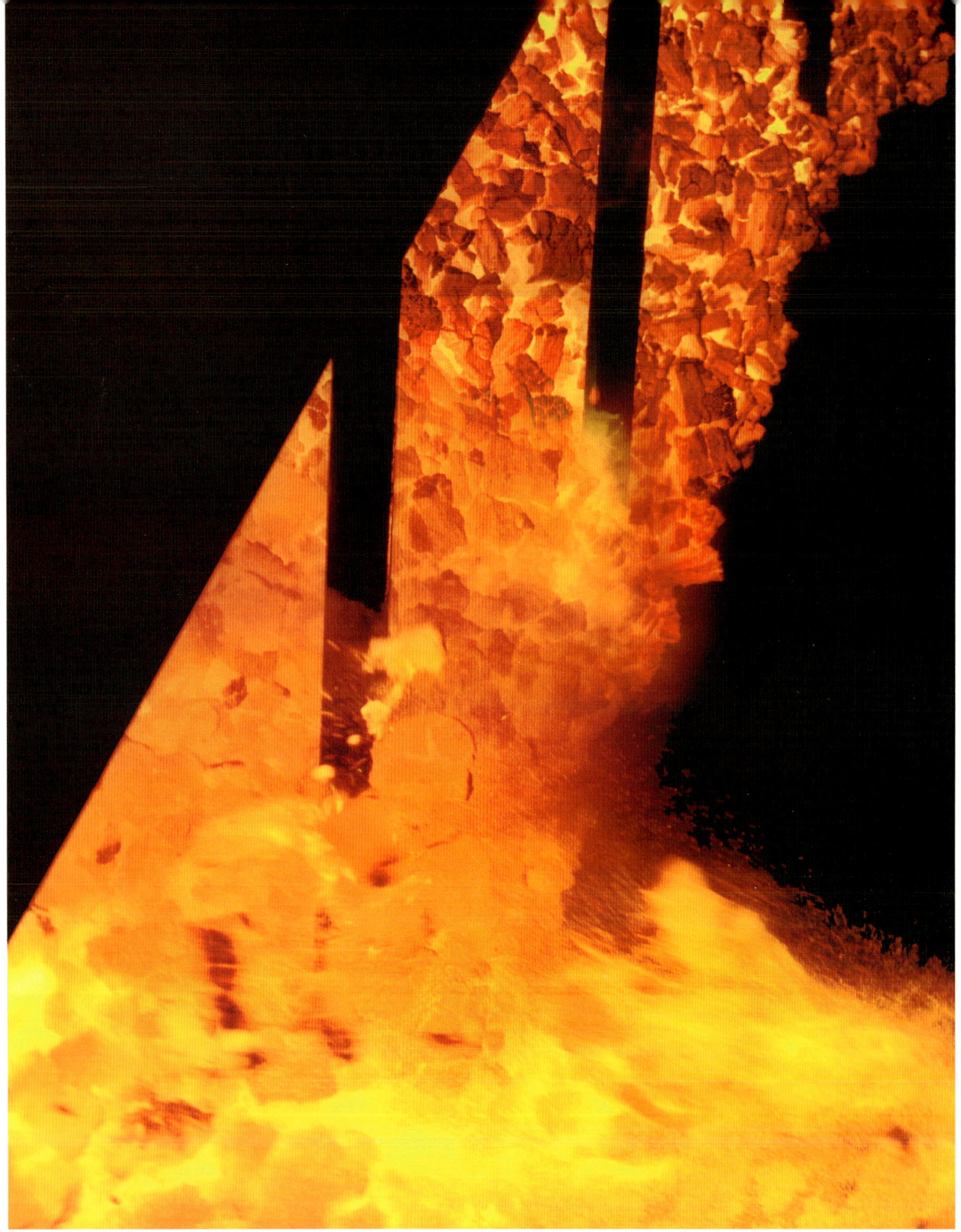

Coke oven with push door off

1

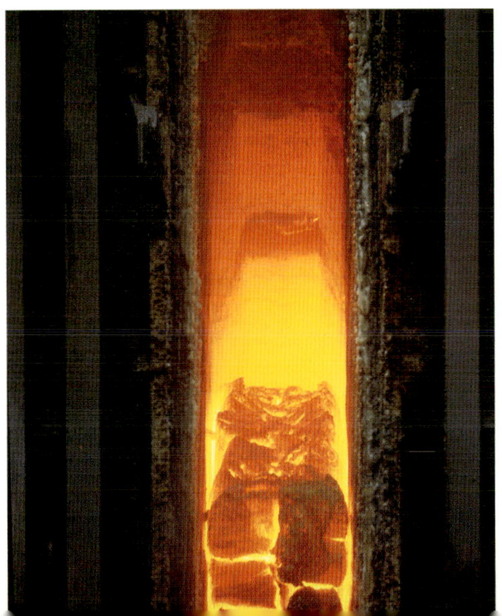

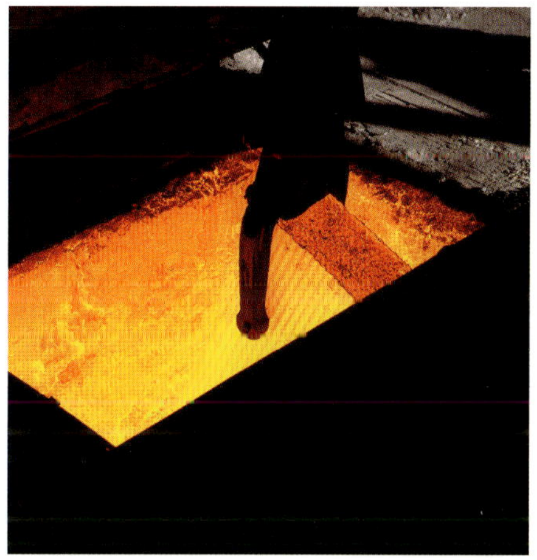

IRON & STEEL MAKING

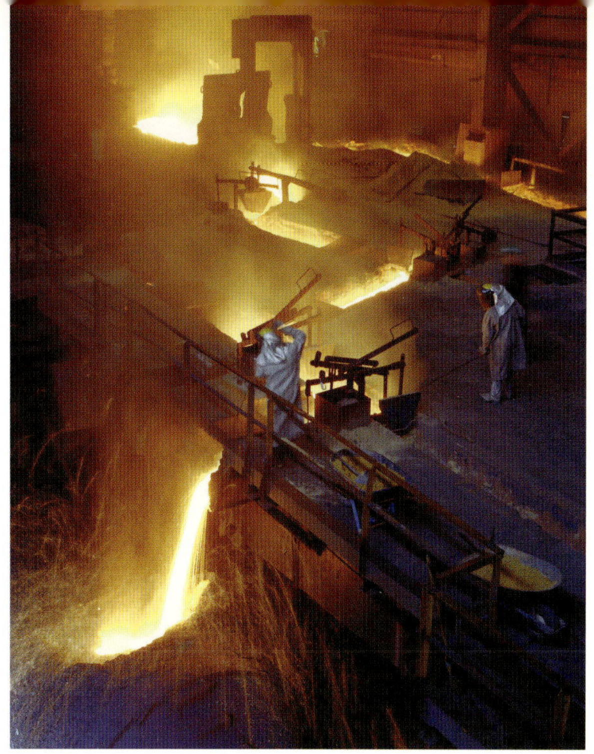

1

Tapping a blast furnace heat

1 In the second step of the fully integrated steelmaking process, the blast furnace transforms coke, iron ore, and limestone into the first form of molten metal. The raw materials are hoisted to the top of the furnace and then deposited in the bell of the furnace where a blast of hot air melts the raw materials into molten iron. In older blast furnaces, as pictured here at the Bethlehem, Pa., plant, iron flowed through open troughs into submarine rail cars positioned under the cast house floor. Newer furnaces constructed in the 1970s such as those at Bethlehem's Burns Harbor, Ind., and Sparrows Point, Md., plants covered their troughs for employee safety and environmental concerns. When the furnace was opened, the majority of the material – the iron – ran through the troughs into the waiting submarine cars while slag, composed of impurities, was diverted to another trough into slag pots, which were also under the furnace's cast-house floor. Even a visitor could tell when the iron run was over: the slag came out with a vengeance and its sulfur content created a pungent smell and choking sensation in the throat and lungs. Slag was sold for construction, manufacturing insulation, and special concrete. A popular use for slag was to prepare roadbeds for paving.

2 The panorama of the Bethlehem, Pa., blast furnaces along the Lehigh River now stand in silent tribute to the hundreds of thousands of men and women who worked at the plant, beginning in 1857 as the Saucona Iron Company and later as the Bethlehem Steel Company. During 1943, the peak year of Bethlehem's production for the Allied forces during World War II, more than 31,000 people worked at the plant to produce important wartime products. Now, the blast furnaces are silent sentinels that will be preserved as part of the Sands BethWorks entertainment complex to be built on 120 acres of former steel plant property along the Lehigh River in South Bethlehem. Blast furnace A, to the far right, was built in 1901 and operated until 1961. Furnaces B through E operated into the 1980s. The last to operate, blast furnace C in the center, closed on November 18, 1995.

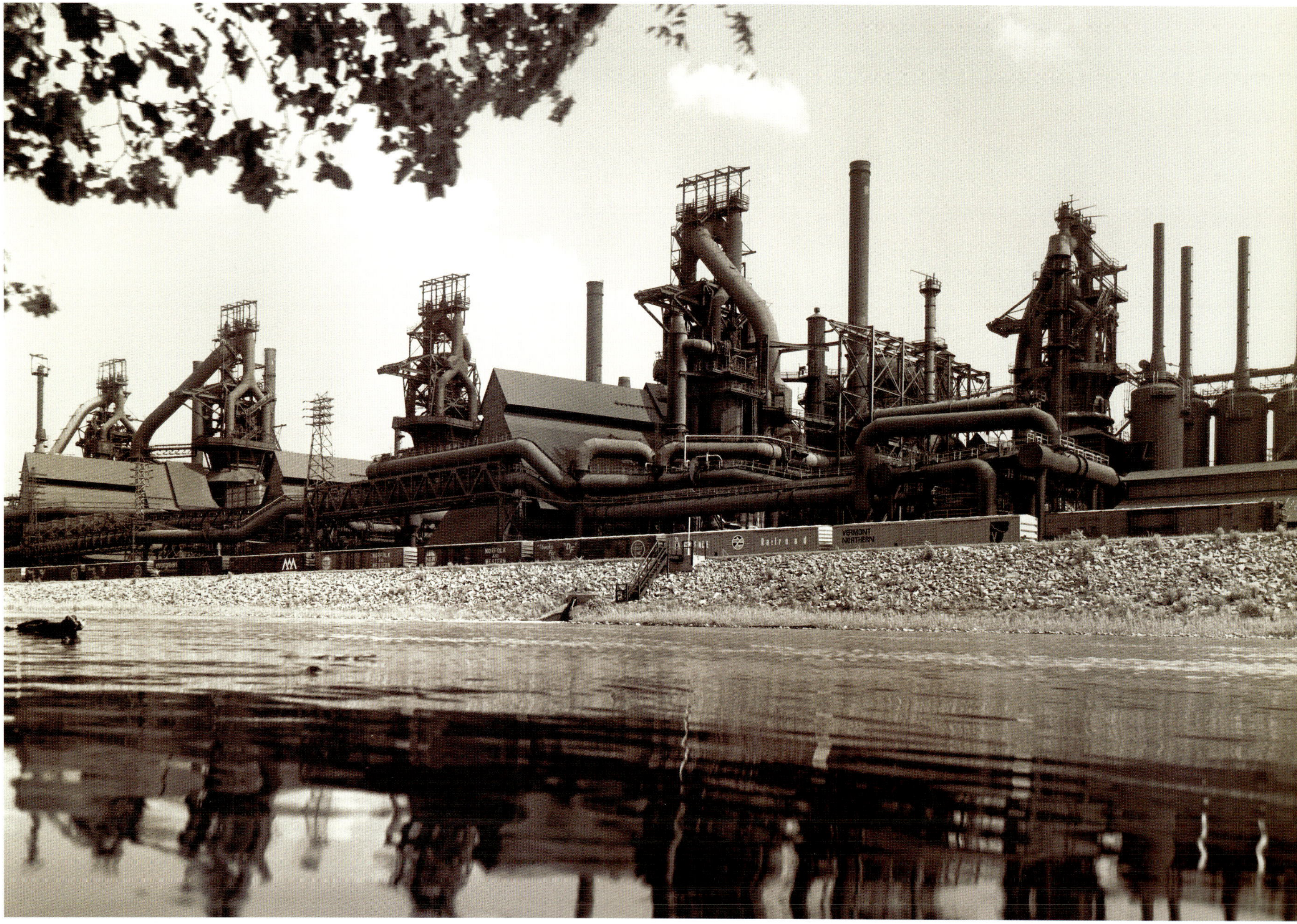

2

Blast furnace row

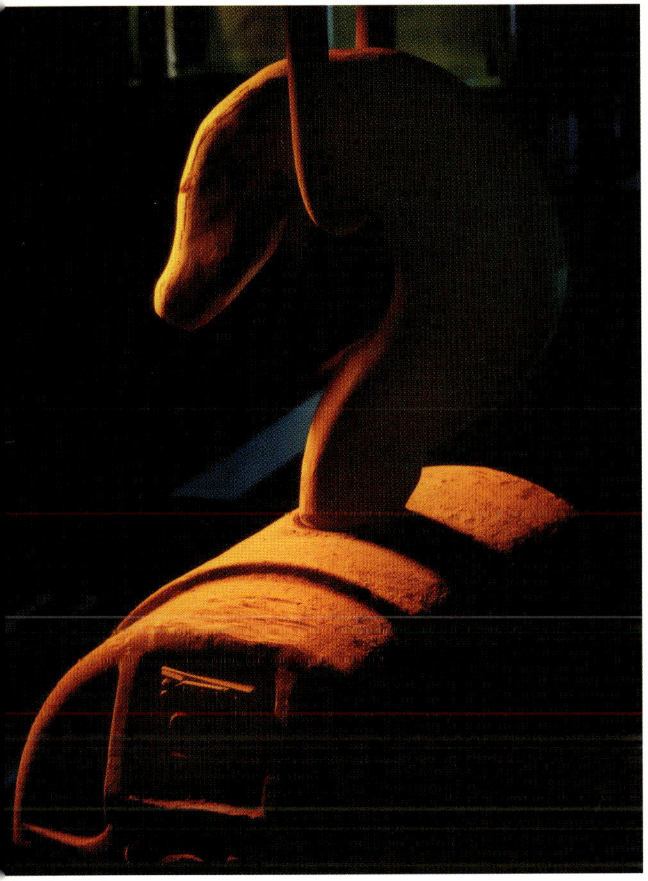

1

Ladle hook

1 The enormity of the scale of steel industry equipment, such as this giant ladle hook, is difficult to comprehend. A ladle weighed 100 tons and could hold 250 or more tons of molten metal. To carry the ladle and its load required cranes with lifting capacities up to 425 tons. While the higher-end steels for automotive and appliance applications are made with the basic oxygen furnace process, many other steel companies have electric arc furnaces that use scrap steel instead of costly and time-consuming raw materials. Bethlehem Steel had electric arc furnaces at its plants in Bethlehem, Steelton, Johnstown, and Coatesville, Pa., as well as at its Los Angeles, Calif., and Seattle, Wash., plants, which were sold in the early 1980s. Bethlehem continued to use electric furnaces for its forged products at its Bethlehem, Pa., plant, along with the open hearth and later basic oxygen furnaces to produce structural shapes – the predominant plant product.

2 The mouth of the basic oxygen furnace silhouettes this steelworker who is taking a metallurgical test of a heat, or batch, of steel produced at the Bethlehem, Pa., plant. The basic oxygen furnace, the current form of fully integrated steelmaking, produces a heat of steel in about 45 minutes. Open hearth furnaces, used from 1888 until 1968, produced the same amount of steel in eight hours. The basic oxygen furnace process was developed by European steelmakers in the 1950s but not embraced by American steelmakers until the 1960s. Bethlehem employed the basic oxygen furnace process at its fully integrated plants in Lackawanna, N.Y.; Sparrows Point, Md.; Burns Harbor, Ind., and Bethlehem, Pa.

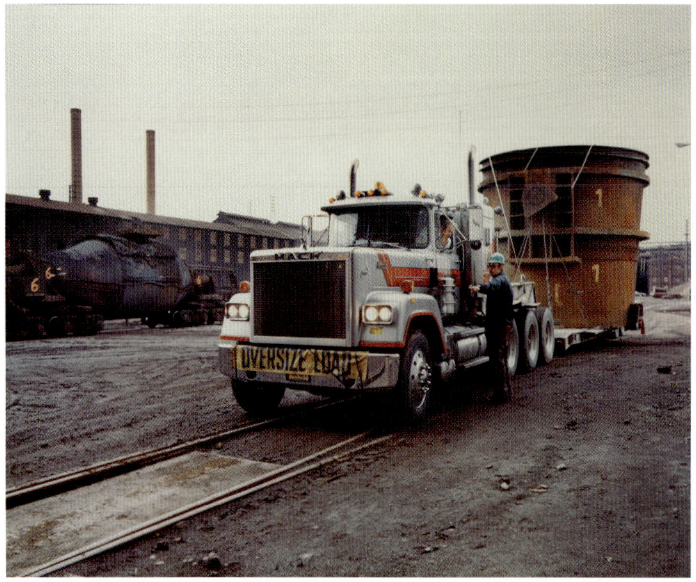

1

A 100 ton load

2

Ingot teeming in Bethlehem

1 The Lackawanna, N.Y., plant was the first fully integrated steel plant in the country to be drastically cut back as the entire domestic steel industry entered a prolonged period of painful closings in the early 1980s. Bethlehem announced the closing of Lackawanna's basic steelmaking and primary operations on December 27, 1982. The announcement became the national news story that signified the beginning of the industry's retrenchment. The $1.15 billion loss for the fourth quarter of 1982, the largest corporate quarterly loss ever recorded by any American company, mostly reflected charges for pensions and health care costs for employees who were laid off, many of whom soon retired. The next five years were the most difficult period of downsizing and prolonged change in the basic fundamentals for steel companies in the United States. As steel plants closed, surplus equipment was transferred within companies or sold to either domestic or international competitors. In the spring of 1984, this ladle, weighing about 100 tons with a capacity of more than 250 tons of molten steel, made a winding journey off highways from the Lackawanna, N.Y., plant to Bethlehem, Pa., on a Mack Truck. Mack Trucks, with headquarters in Allentown, Pa., used this shot for its 1985 product calendar. To the left of the photo is a submarine rail car that transported molten iron from the blast furnace to the basic oxygen furnace shop where this ladle was used.

2 Ingots were the first form of solid steel. After being created in the basic oxygen furnace shop, the steel was poured from the furnace vessel into another ladle that traversed a crane runway in the teeming aisle. The ladle, with a gate at its bottom, filled ingot molds that traveled on narrow-gauge tracks. When full, the molds were taken to a building where they were stripped from the ingots. The molds for all of Bethlehem's plants, and some molds for trade customers, were made at the plant in Bethlehem, Pa. As steelmakers, including Bethlehem, increasingly invested in the time and cost saving method of continuous casting to produce the first solid form of steel, demand for ingot molds steadily declined. The ingot mold foundry was closed in November 1995 coinciding with the end of basic steelmaking and primary operations at the Bethlehem, Pa. plant.

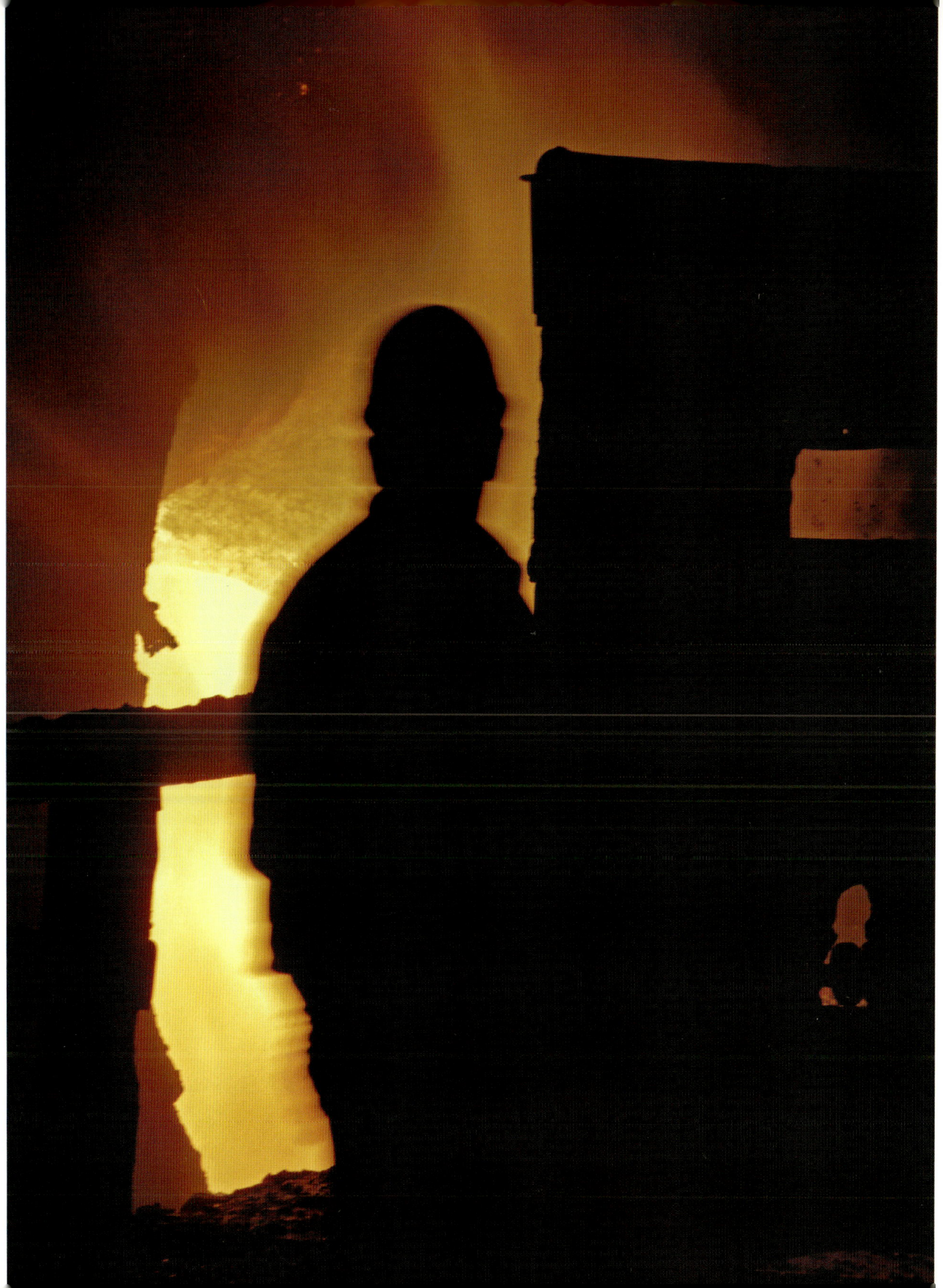

2

Testing a heat, BOF

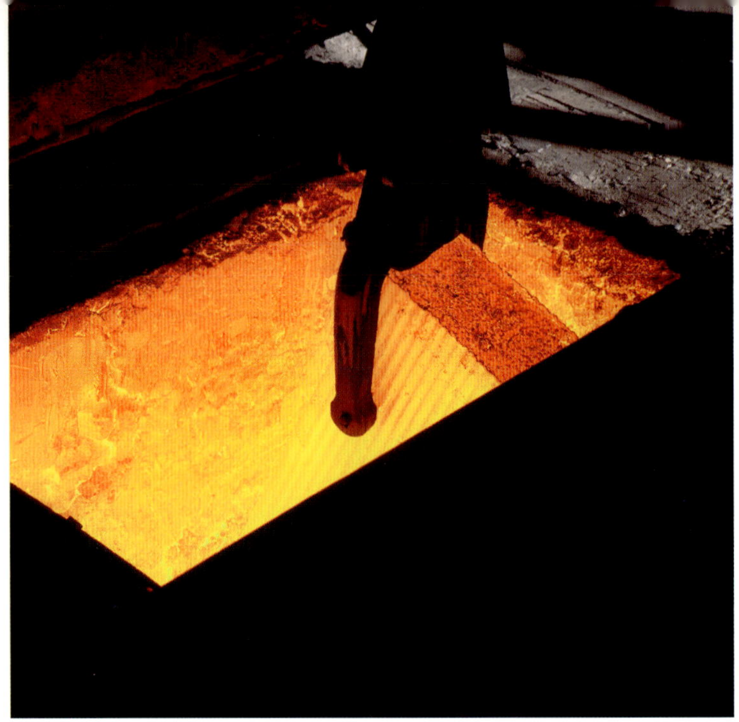

2

Soaking up the 2,350° F heat in Bethlehem

1 By the 1980s, ingot teeming and stripping had become an inefficient method of producing the first solid form of steel. Once formed, the ingot had to be heated again before it could be rolled to produce one of three forms of semi-finished steel – billets, blooms, or slabs – for the mills. Billets were used for thinner, longer products such as bars, rods, wire, and reinforcing bars. Blooms created longer but thicker products such as rails and structural shapes. Slabs were rolled into plates and sheet steels. These ingots, on narrow gauge track buggies, produced at the Bethlehem, Pa., plant, were destined for the plant's structural mills, which accounted for about 90 percent of the finished product shipped by that facility. (Ten percent of shipped product was forged or pounded, instead of rolled or squeezed, into shape.) In 1984, the Bethlehem plant employees, instead of outside contractors, built a new ingot stripping facility adjacent to the basic oxygen furnace. The facility was completed on time and within budget, and was a great source of pride for plant employees.

2 These ingots, destined to become structural shapes for buildings and bridges, were being readied to roll into blooms at the Bethlehem, Pa., plant. Before the advent of the continuous casters, ingots were produced and then broken down into smaller sizes in primary mills. Older plants, such as the Bethlehem, Pa., plant, which were not considered "core" businesses, pushed on with the ingot process. After the ingots were cooled and stripped of their molds, they were then reheated to 2,350° F in soaking pits – large, refractory brick-lined boxes with heat to make the ingots pliable enough to enter the rolling mills. These soaking pits were rebuilt in the early 1980s. During their construction, a wall collapsed, killing three plant employees.

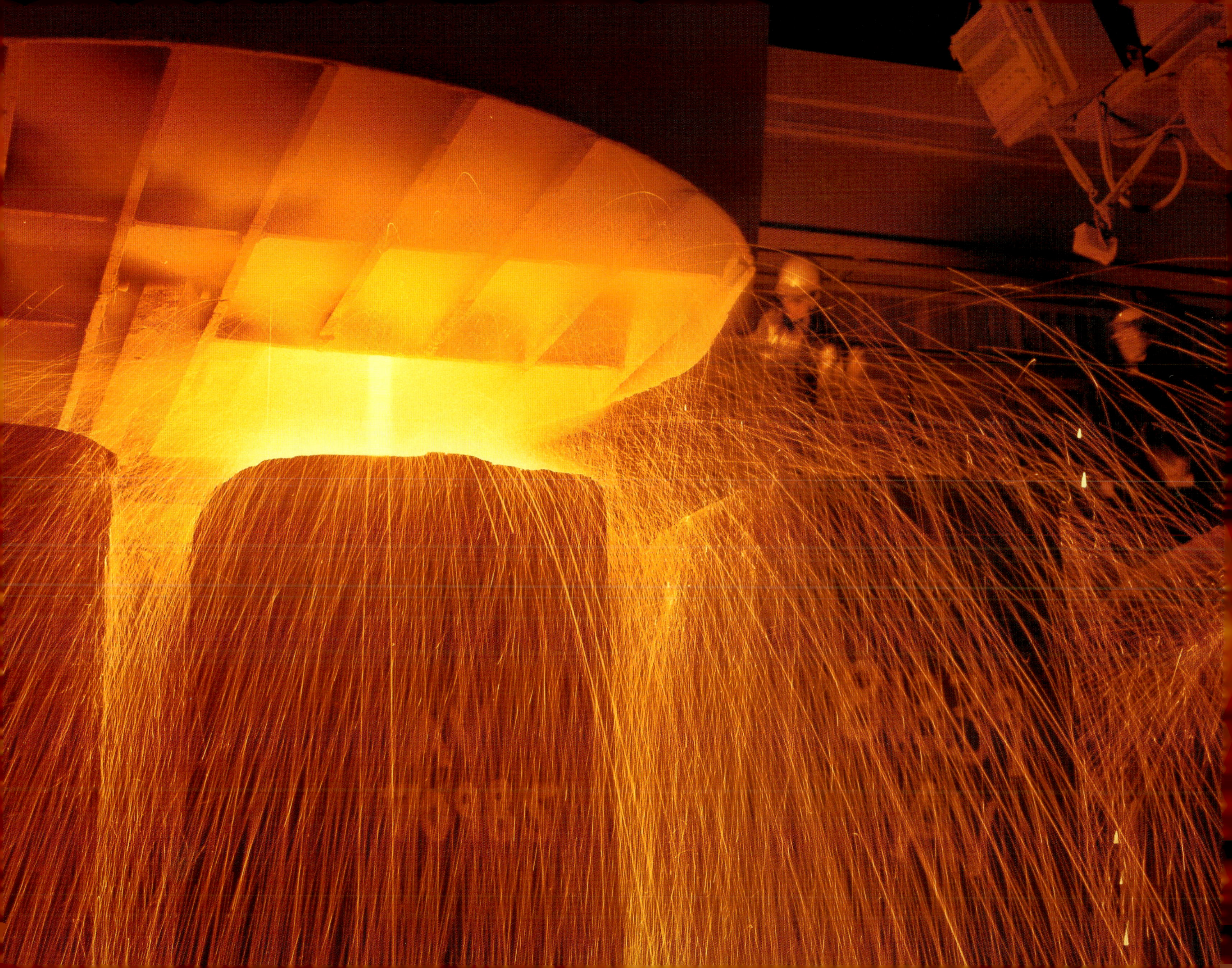

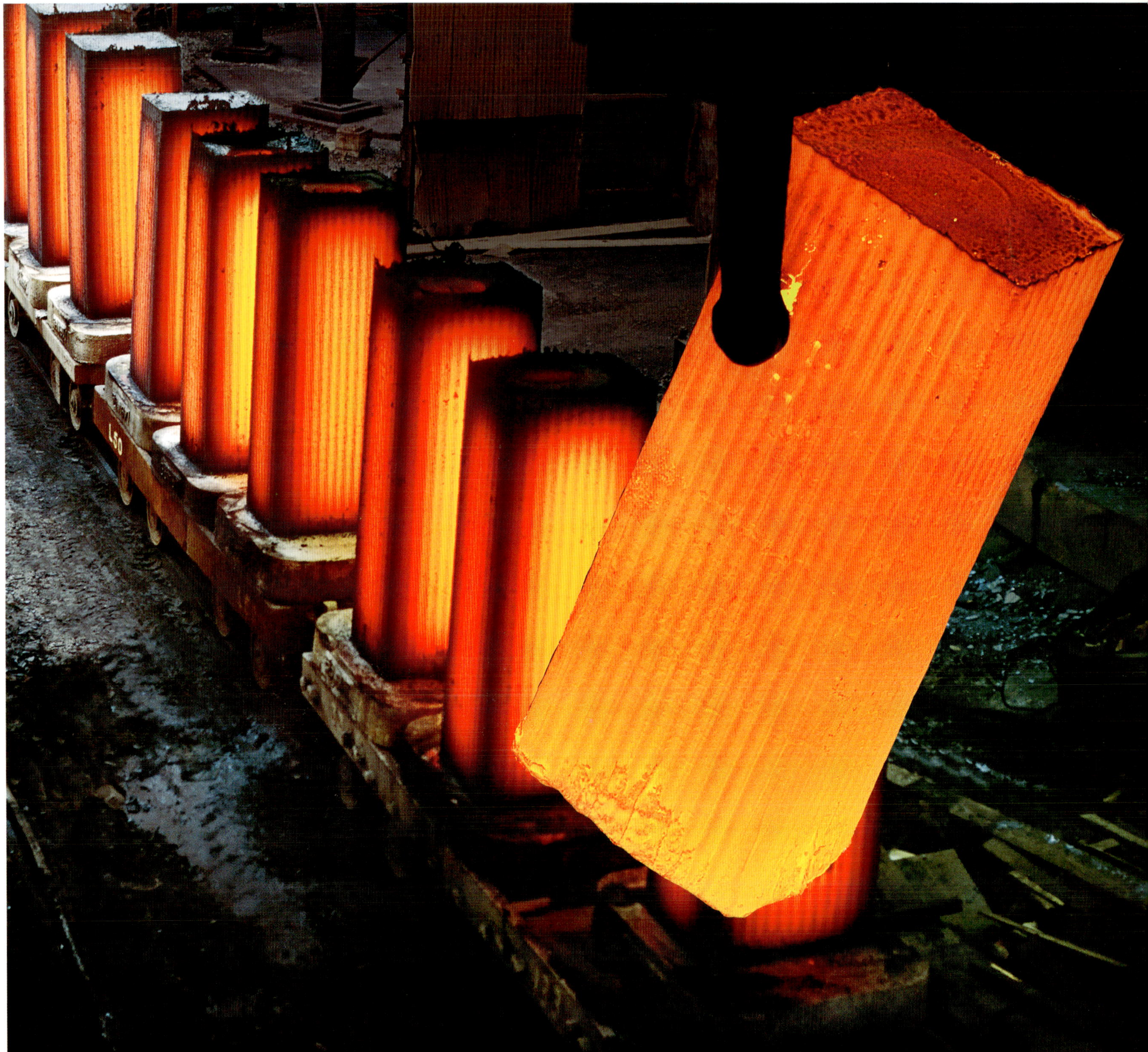

1

Ingots on narrow gauge buggies

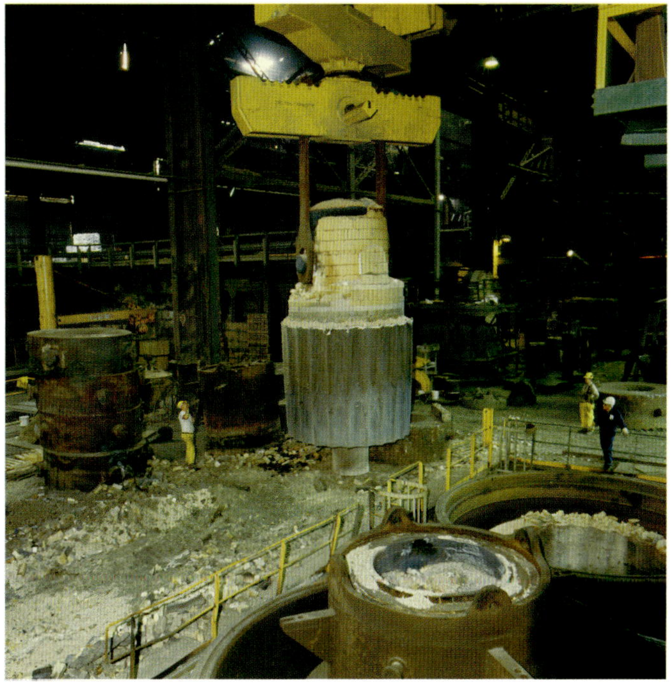

Giant electric furnace ingot in Steelton

Forged products were long part of the Bethlehem, Pa., plant, dating back to the 1880s when the forging and final machining capabilities of the plant convinced the U.S. Navy to retool its fleet from wood to metal. Forged products helped America win the Spanish American War in 1898 and both World Wars in the first half of the 20th century. In 1995, at nearly the same time that basic steelmaking at the Bethlehem plant ended, electric steelmaking in Bethlehem also was discontinued. Just a year earlier in 1994, Bethlehem had installed a new 150-ton DC electric arc furnace with vacuum degassing and ladle refining at its electric furnace shop in Steelton, Pa., 90 miles from the Bethlehem plant's forging operations. To maximize efficiency and to satisfy the larger-product needs of the U.S. defense and energy industries, Steelton created 320-ton ingots in its electric furnace shop for shipment to Bethlehem. The ingots traveled in insulated rail cars to maintain their temperature and minimize reheating time before forging.

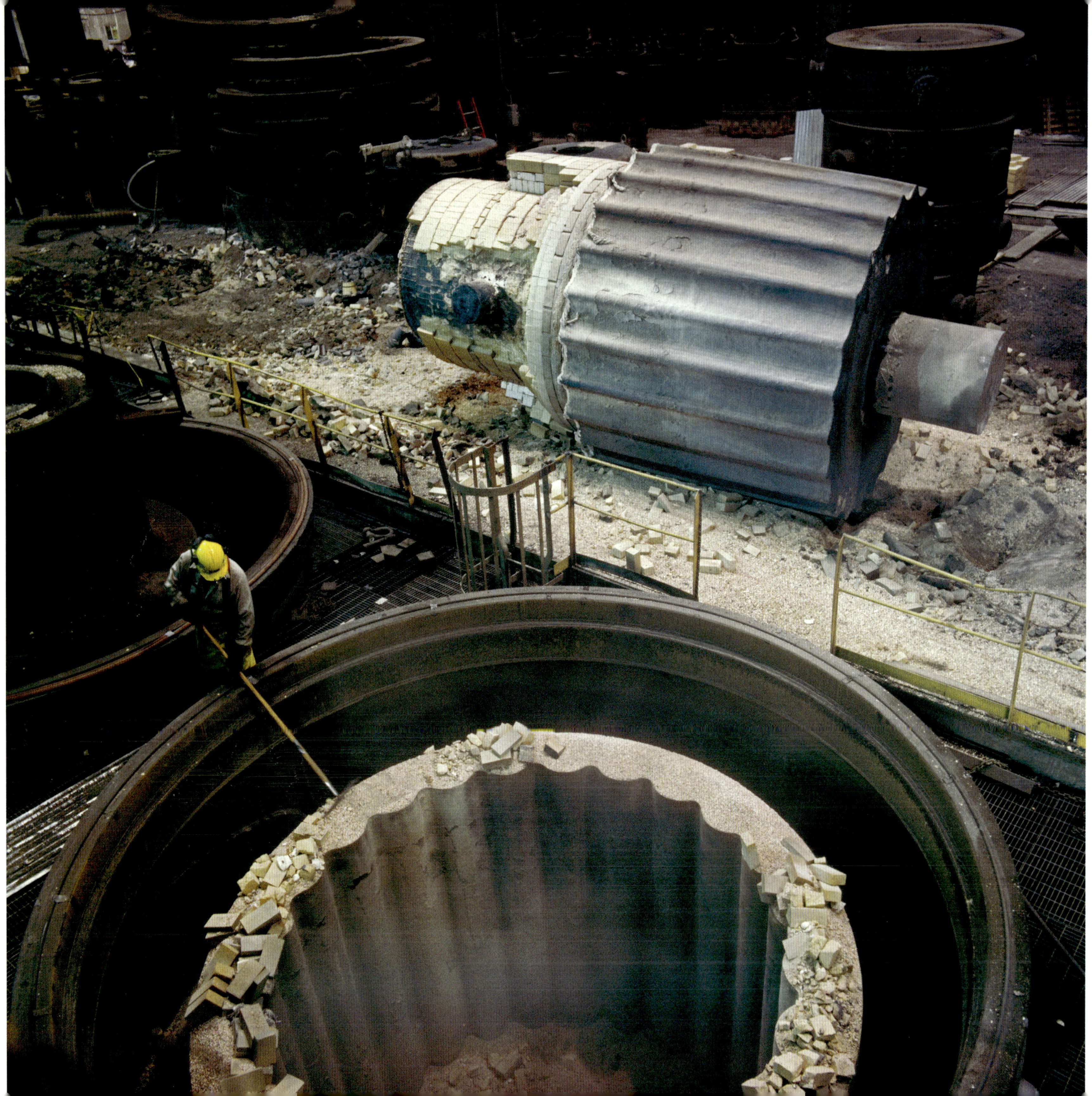

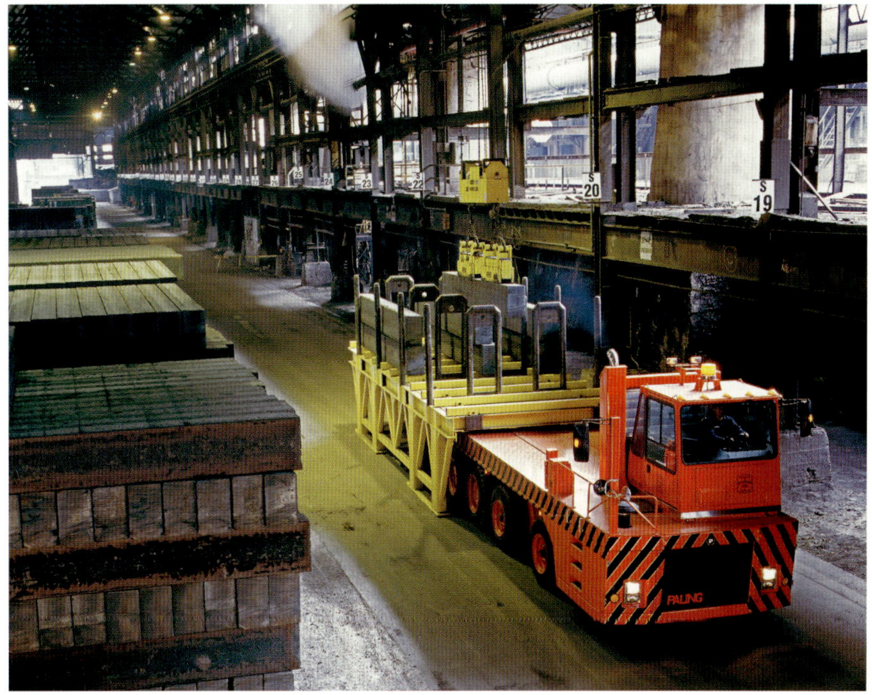

1 Loading steel blooms

1 To supply the combination rolling mill with steel, Bethlehem Steel shipped blooms from its continuous caster in Steelton, Pa., which was installed in 1983, to the Bethlehem, Pa., plant. Upon arrival by rail in Bethlehem, the blooms were loaded onto these pallet carriers that stored the blooms in an abandoned open hearth building close to the combination mill. The efficiencies of maximum bloom rolling at Steelton combined with minimal handling at Bethlehem were supposed to be profitable and logical, but they were neither. November 19, 1995, the day the Bethlehem furnaces were silenced and the end of steelmaking in Bethlehem, was a traumatic event for the city of its name. More than 1,800 employees went home that month, and an equal number remained at the coke ovens and combination rolling mill.

2 Bethlehem's Burns Harbor, Ind., plant opened in 1964 with much fanfare as a finishing facility and the newest fully integrated steel plant in the United States. In the early 1970s, Bethlehem installed a three-vessel basic oxygen furnace and, later that decade, a continuous caster. Continuous casters were expensive – about $250 million each – but essential to survival from the onslaught of foreign competition and the leaner, meaner so-called minimills. Both integrated steel mills and minimills used continuous casters and electric arc furnaces fed by inexpensive scrap steel instead of more costly and time-intensive raw materials. In the 1980s, as the industry was reeling from unprecedented losses and the closing of inefficient mills, steelmakers started to look to the future. Bethlehem invested in a second caster at Burns Harbor and added one each at Sparrows Point, Md., and Steelton, Pa. Those facilities without casters, most notably Lackawanna, N.Y., and Bethlehem, Pa., believed their days were numbered. Lackawanna was mostly closed in 1983, but Bethlehem hung on into the mid 1990s.

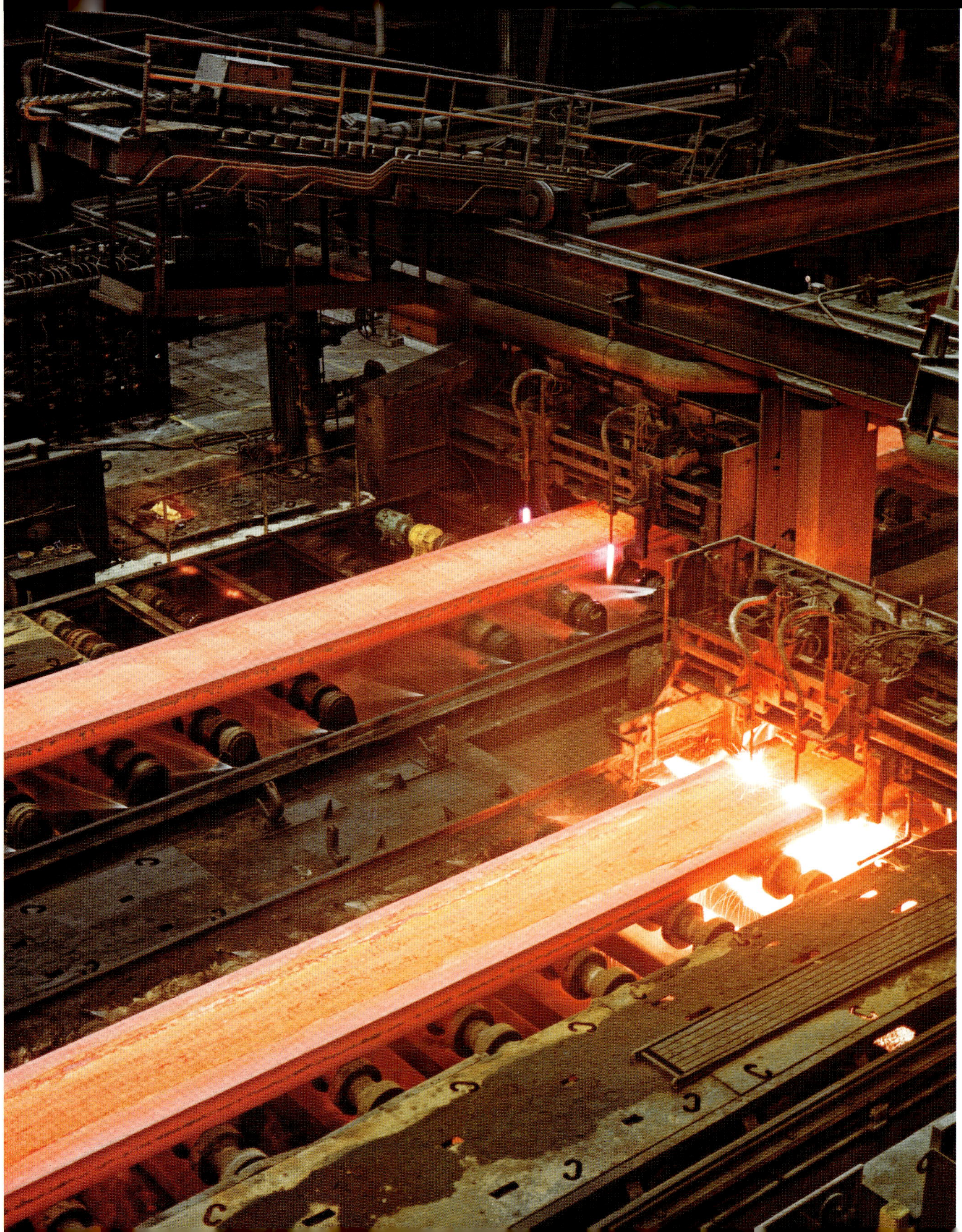

2

Flame cutting continuous castings into slabs

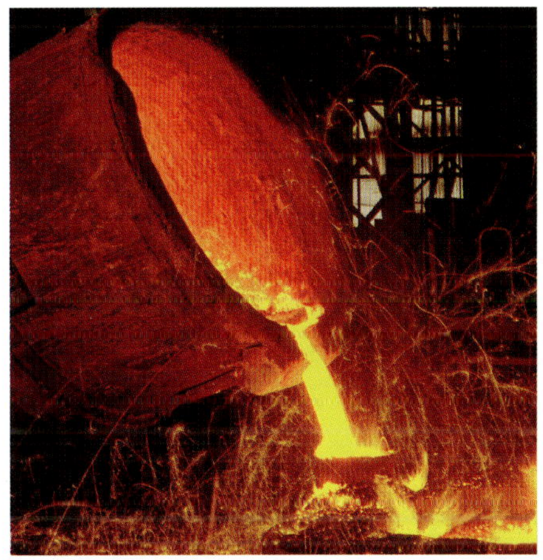

Bethlehem's defensive moves to stave off bankruptcy court protection, which was considered and nearly enacted in 1986, lasted from 1981 through 1987. Inefficient facilities were closed, including nearly all of the Lackawanna, N.Y., plant, the first fully integrated steel plant in the nation to be shuttered that decade. Tens of thousands of Bethlehem employees were laid off. Wages and benefits were slashed. Product lines were sold off or discontinued.

During those tumultuous and painful years, however, Bethlehem began to focus sharply on the future, which was clearly in its highest-value products of light-flat-rolled steel produced in Maryland and Indiana. Those steels were highly sought after by automakers with U.S. operations. Research and development by hundreds of highly educated metallurgists at the firm's Homer Research Laboratories helped put the company's products back in favor with automotive, appliance, and metal-building companies. Believing its vision was "to be the premier steel company," Bethlehem's painful choices in the 1980s let the company live to enter the 1990s.

Bethlehem's market share eroded further in the next decade as imports and the minimills increased their share. Nearly all mills at the Johnstown and Bethlehem, Pa., plants were closed in the early to mid 1990s. In the meantime, Bethlehem Steel invested heavily in the mills that made light-flat-rolled steel for automobiles, appliances, and metal buildings. The operations at Burns Harbor, Ind., and Sparrows Point, Md., became the nucleus of the corporation. In 1998, Bethlehem completed the acquisition of Lukens Inc., the oldest, continuously operating steel company in the United States, and became the largest and most versatile domestic plate producer. The transaction signaled the beginning of an industry-wide consolidation that would, eventually, lead to the demise of Bethlehem Steel.

As the corporation sold assets in the 1970s through the 1990s, its laid-off employees increasingly became pensioners who received company-paid health care and defined pension plan benefits. In 1983, the number of each population was equal. By 2001, about 12,000 employees supported 74,000 retirees and 120,000 health care recipients. Through most of 2001, the stock market was eroding, and the result of the September 11th terrorist attacks sent the stock market into a tremendous free fall, escalating Bethlehem's underfunded pension obligation to about $2 billion. On October 15, 2001, Bethlehem filed for Chapter 11 bankruptcy court protection. In May 2003, all of Bethlehem's ongoing and idle assets were sold to International Steel Group, a company that was composed primarily of the assets of the defunct LTV Steel. "Bessie," the darling of Wall Street that produced the steel to build, transport, and defend America, had ceased to exist.

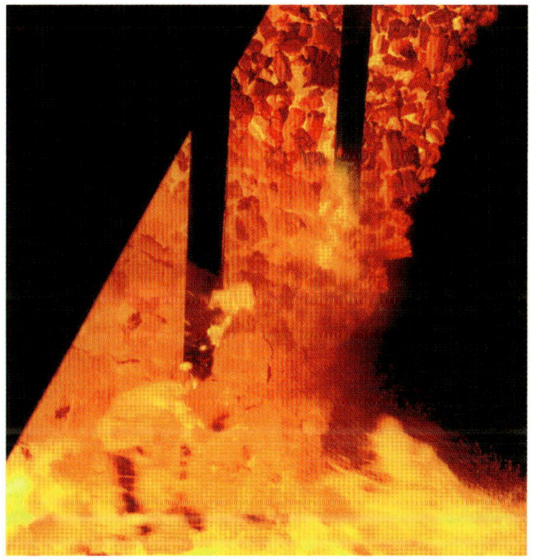

INTRODUCTION

Bethlehem Steel Corporation
1904-2003

For nearly its entire existence – from its incorporation in December 1904 until its sale in May 2003 – Bethlehem Steel Corporation was the second-largest steel company in the United States of America. During its life, it also was a large, worldwide company engaged in the extraction of raw materials and, in the years of World War II, the world's largest shipbuilder.

In 1906, Bethlehem Steel invested in new technology – the continuous rolling of wide-flange structural shapes – that, when operational in 1908, ushered in the age of the sky-scraper and the long-span bridge. The structural products of the Bethlehem, Pa., plant became the symbol for the company and spawned a nationwide series of facilities employing millions of people during the company's nearly 99-year life.

The early and tremendous successes of Bethlehem, nick-named Bessie, as a shipbuilder, forging plant, and far-flung series of steel plants earned the company a position on the Dow Jones Industrial Index in 1918. It was one of the nation's most favored stocks traded in New York City from the 1910s through the 1970s.

In the heydays of the 1910s through the 1960s, Bethlehem Steel produced and erected the steel for American iconic structures such as New York City's Madison Square Garden, Lincoln Tunnel, Waldorf Astoria Hotel, Chrysler Building, and George Washington Bridge. Elsewhere in the United States, employees made and fabricated the steel for San Francisco's Golden Gate Bridge; Chicago's Merchandise Mart; the Virginia Bay Bridge Tunnel; Los Angeles' City Hall; and Washington, D.C.'s U.S. Supreme Court and U.S. Congress' Rayburn buildings. During the 1910s and 1920s, Bethlehem also bought steelmaking facilities and coal reserves in Maryland, Pennsylvania, New York, West Virginia, California, and Washington to further expand its empire and its product lines.

During World War II, Bethlehem Steel was the largest shipbuilder in the world and employed nearly 300,000 people at its plants, shipyards, and mines to supply steel to the Allied war effort. Its vast and numerous shipyards contributed more than a ship a day (just under 1,200) to defend America and defeat the enemies in the European and Pacific arenas, and the plants produced millions of tons of steel for every imaginable military product.

After World War II, America's nearly insatiable appetite for steel kept Bethlehem Steel's mills across the country bustling and highly profitable. The nation's infrastructure was expanded to include higher buildings in our cities, new roadways to carry traffic, and bridges to safely span America's waterways, giving way to the explosive growth of the suburbs. Production of light-flat-rolled steel grew as Americans took to the newly constructed roads in their new American cars.

But the chinks were already forming in Bethlehem Steel's armor during the heady days of the 1950s. According to *Fortune* magazine, six of the ten highest paid executives (salary and bonus) in all of corporate America were in the upper echelon of Bethlehem Steel. The 116-day strike in 1959 by the United Steelworkers of America against the "Big Eight" steel companies, including Bethlehem, shuttered all domestic steelmaking operations. The world was watching, and the door to the lucrative American steel market was flung wide open. In marched the off-shore producers, the Japanese in particular, to fill the void. Bethlehem, known as the U.S. "supermarket of steel," was in for a tough fight after foreign steel made its presence known in the U.S. marketplace.

Despite the presence of foreign steel in the U.S. marketplace, the 1960s and 1970s were mostly profitable, fueling huge capital investments. Most of these expenditures were for new processes (namely basic oxygen furnaces in the 1960s and one continuous caster in the 1970s) merely to catch up to Bethlehem's foreign competition while others expanded growing product lines for corrosion- and dent-resistant light-flat-rolled steel for automotive and appliance makers. Environmental quality control expenditures were mandated by the federal Environmental Protection Agency, which was commissioned in 1970. Air and water quality control systems added to operating costs and eroded profitability. At the same time, foreign steel companies and new, low-cost minimills were nipping at the heels of Bethlehem and its kindred fully integrated domestic steel producers.

In 1977, Bethlehem Steel experienced one disaster after another. In January, a blizzard – with up to 14 feet of snow on the shore of Lake Erie – nearly crippled the Lackawanna, N.Y., plant. In February, a fire in the Cambria, Pa., coal mine north of Johnstown, Pa., raged out of control and curtailed production for two months. In July, the Johnstown plant was heavily damaged when more than one foot of rain fell on the city, causing rivers to swell and dams to break. In the fall, on a day that became known as Black Friday, Bethlehem Steel reported its first corporate quarterly loss in 43 years, closed duplicate rolling mills at marginal plants, and laid off white-collared employees in the headquarters. But the bad news was not over; in December, two oil tankers owned by Bethlehem collided in a remote shipping lane off the coast of Africa.

The bad news of 1977 was just a prelude to the turmoil of restructuring the entire Bethlehem Steel Corporation that began in 1981. A nationwide recession resulted in fewer capital outlays for durable goods such as cars and appliances. The sky-high interest rates put building projects on hold. Foreign competition was ruthless, and the minimills were continuing their quest to gain inroads into – and eventually capture – the market for long products such as rebar, bars, rods, wire, and, eventually, structural shapes.

RAW MATERIALS

1

Mine roof bolting

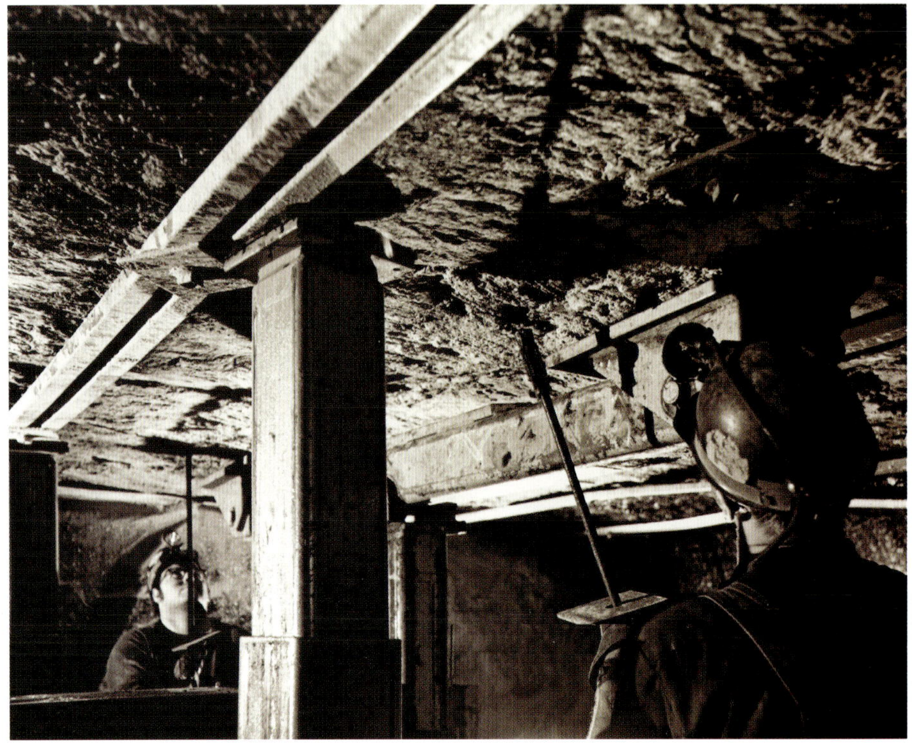

1 Bethlehem Steel mined soft bituminous coal in underground mines for its fully integrated steel plants in the mid-western and eastern United States. Underground mines were either slope mines where workers entered in a cart, called a "man trip" that gradually descended into the mine from ground level, or vertical-descent mines where workers descended up to 500 feet by elevator. Roof bolts, produced at Bethlehem's Industrial Fastener Division in Lebanon, Pa., were an important component of maintaining the roof of the mine while the employees removed the soft coal. The corporation also owned an open-pit mine, named Panther Valley, in Tamaqua, Pa., to harvest hard anthracite coal.

2 The key raw material for making steel is iron ore, or taconite, which is extracted from two open, or surface, mines – one in Hibbing, Minn., operated by Cleveland Cliffs and one in Labrador City, Newfoundland, Canada, operated by the Iron Ore Company of Canada (IOC). Once mined, the ore was transported in these huge trucks to on-site pelletizing plants. Ore from Hibbing was loaded onto ships that sailed the Great Lakes to Bethlehem's Burns Harbor, Ind., plant on the southern shore of Lake Michigan while IOC's ore was loaded into barges that sailed the St. Lawrence Seaway into the Atlantic to the Port of Philadelphia. From Philadelphia, the Newfoundland ore was transferred into rail cars and shipped to its final destination, primarily Bethlehem's plants in Sparrows Point, Md., and Bethlehem, Pa. Slight spillage along train tracks prompted local residents to call the small round pellets "train poop." Bethlehem sold its interests in IOC in the late 1990s and its partial ownership of Cleveland Cliffs in May 2003.

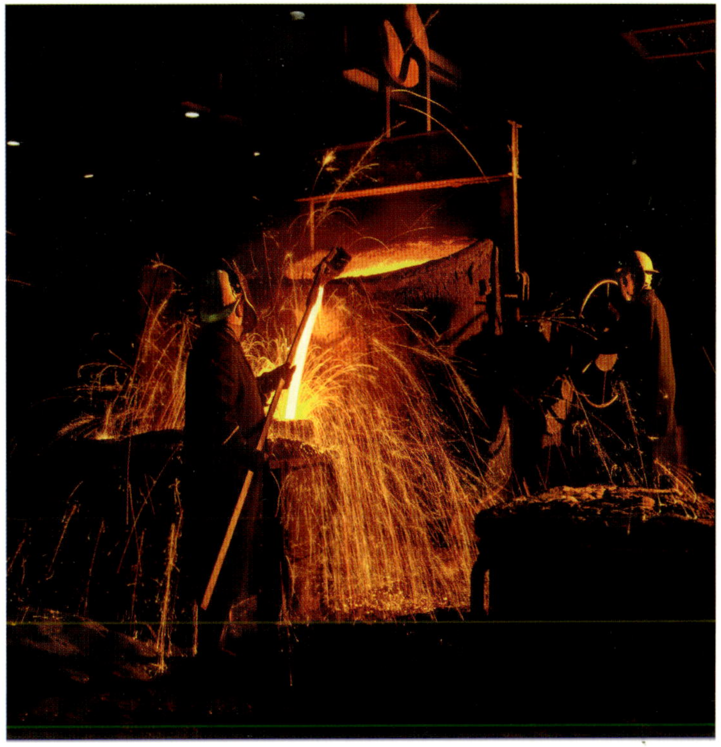

2

Removing slag from ladle

1 Bethlehem Steel's foundries were smaller versions of the teeming process at the exit end of the steelmaking shop. Small amounts of iron and steel were sent to each respective foundry where the hot metal was cast into molds for steel plant products and for specialized products for other industries. The small scale of the foundry operations is reminiscent of the early days of hot metal production seen at living history sites such as Saugus, Mass., which is part of the National Park Service and revered as the birthplace of metal making in the United States.

2 Once the hot metal arrived, it was poured into a mold that was made by packing sand around the pattern of the desired shape. After the metal solidified, the mold was moved to a shakeout station that separated the casting from the sand. Once all sand had been removed from the newly formed part, its rough edges were smoothed. To contain any spillage, Bethlehem teemed large foundry products in 10-feet deep casting pits. Patterns for foundry products were made out of wood or metal and stored in a nearby pattern shop. While most patterns were destroyed after the Bethlehem plant's iron foundry was closed in 1995, some mysteriously found their way out of the plant and have been incorporated by artisans into small furniture and other conversation pieces. Seeing the patterns destroyed brought tears to the eyes of photographers who were capturing images for archival purposes.

FOUNDRIES

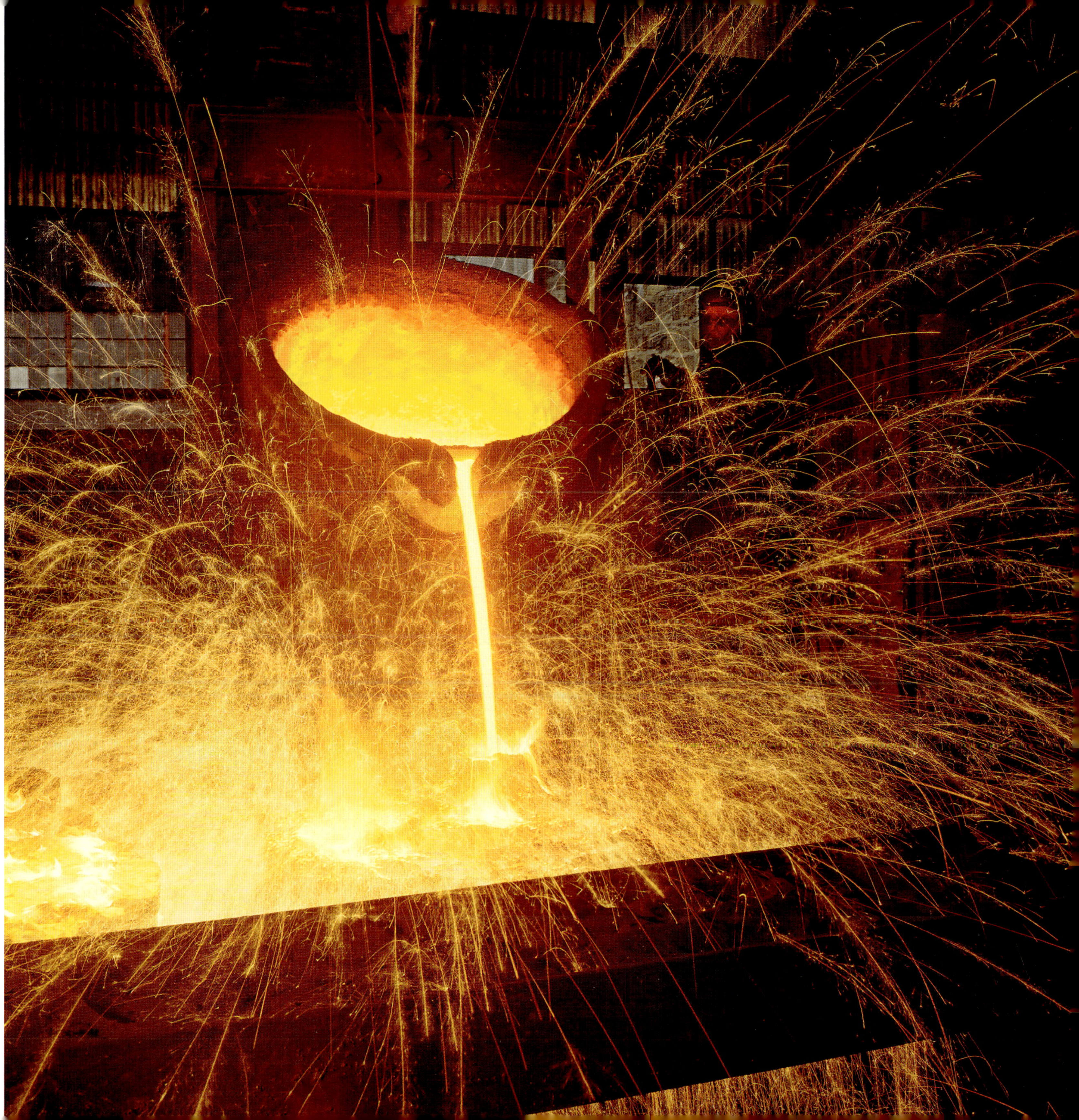

1

Casting in iron foundry

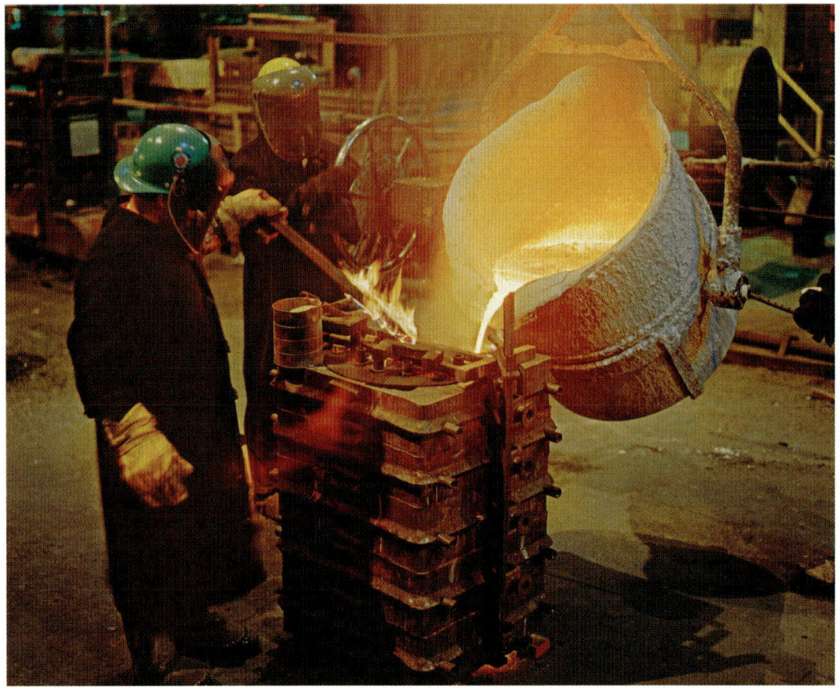

1

Casting small parts

1 The steel foundry operated on the same principles as the iron foundry – using hot metal within patterns to form final pieces. The Bethlehem, Pa., plant's steel foundry was the most versatile of similar foundries at other corporate plants, and received its steel from the electric furnace shop that provided metal to the forging division. After World War II, Bethlehem's foundry division included the iron foundry, steel foundry, brass foundry, ingot mold foundry, roll foundry, No. 10 roll machine shop, and the pattern shop. The Bethlehem, Pa., plant served as the nucleus for foundry products consumed throughout the corporation. With the widespread use of continuous casters, demand for ingot molds drastically declined.

2 The iron foundry at the Bethlehem, Pa., plant was located in the second oldest plant building. First opened in 1873, the building served as the site of the early steelmaking and rail rolling. The cruciform-shaped building was constructed of the nearby and plentiful Pennsylvania field stone to make it fire resistant. Huge arched windows allowed the maximum amount of daylight and ventilation needed in the early days of operation prior to widespread electrification. When the building first opened, it was attached to another, identical building. As the plant grew and as technology changed, sometimes the exterior of the original buildings changed too. In the case of the later years of the iron foundry building, little of the field stone exterior was visible from the outside because smaller structures had been added adjacent to the original façade. Only after the plant was closed and selective structures demolished to make way for redevelopment was the beautiful stonework of the original building visible once again.

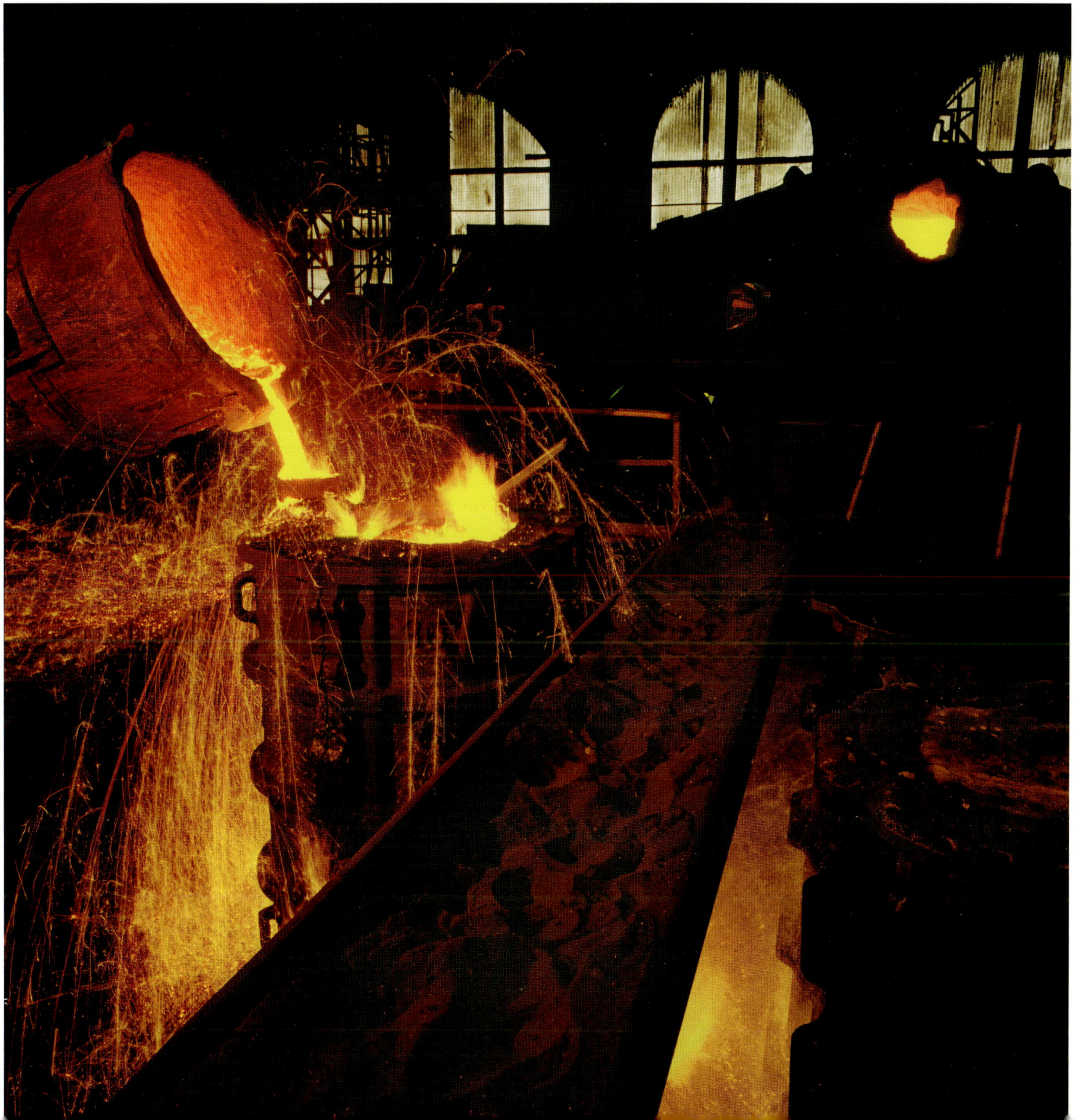

2

Casting pit

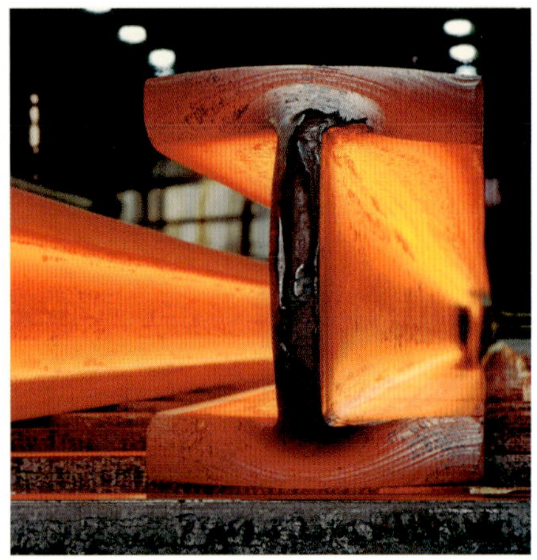

ROLLED PRODUCTS

The iconic symbol of Bethlehem Steel

The legendary I-beam, introduced by Bethlehem to the marketplace in late 1908, started to make its presence known the following year when Adam Gimbel built his flagship midtown Manhattan department store with 12,000 tons of steel. The Gimbel Brothers building was the first to use the new beams on a large scale, and during construction the store was donned with a large banner proclaiming it was "Built the Bethlehem Way." The beam, also known in its early days as the Bethlehem Beam, provided an efficient and economical method of high-rise construction. Previously, several steel sections had to be riveted together to provide the structural stability needed for successively higher floors. The old process was time consuming and expensive, which made wide-flange beams an instant favorite with building contractors and owners looking to create more square footage for less money.

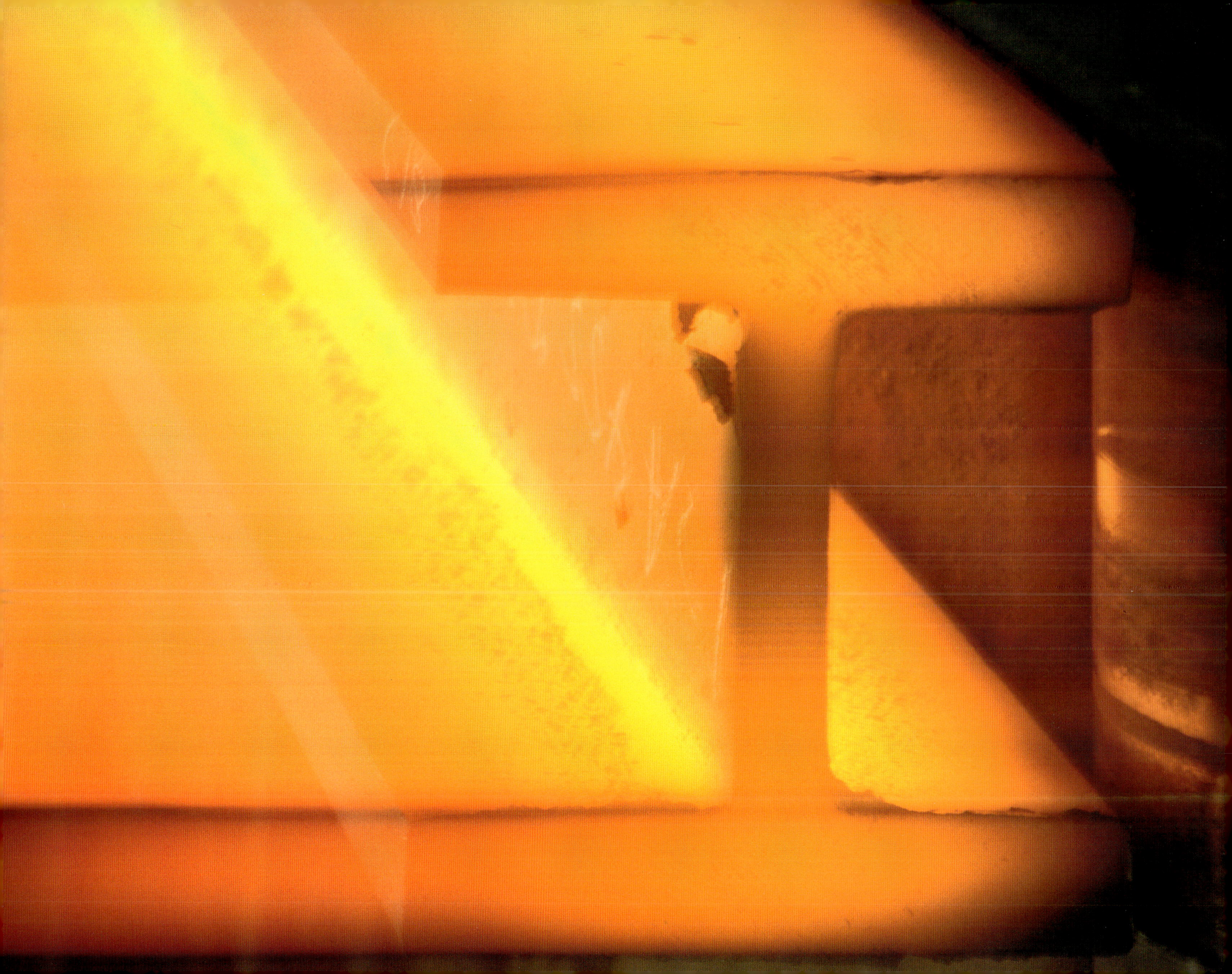

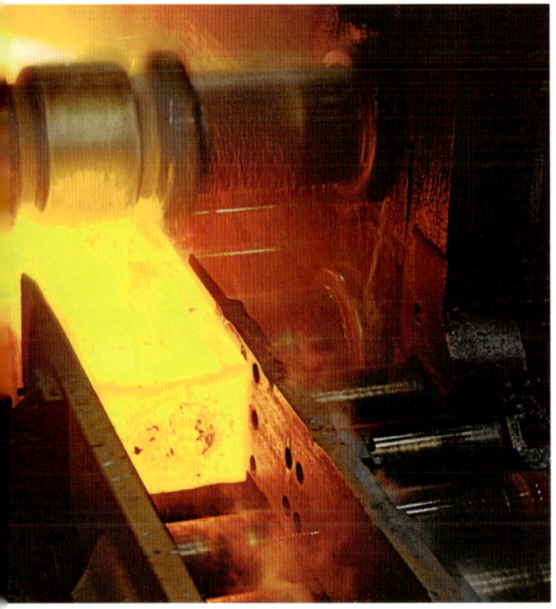

1

Ingot breakdown

1 Bethlehem Steel's most famous product was the structural shape known as the "wide-flange beam." It was adopted as the company's symbol and was recognized internationally. Even after Bethlehem exited the structural products market in 1997, market research indicated that the wide-flange beam was synonymous with the company, prompting corporate officials to keep the symbol. After being teemed and stripped, the ingots entered a blooming mill to reduce their size for rolling in the Bethlehem, Pa., plant's rolling mills for a variety of wide-flange shapes. Although the Bethlehem plant was best known for being the corporation's largest provider of structural shapes, it was not the only plant in the Bethlehem system to roll that product line. The Lackawanna, N.Y., plant produced wide-flange beams until 1977, when all structural-shape production was consolidated at the Bethlehem, Pa., plant.

2 In the blooming mill, the beam starts to take shape. The force of the rolls on the steel and the turning of the ingot create the form needed to enter the finishing mills. Water washes away impurities, known as scale, which form on the surface during rolling. Ingots enter the blooming mills as rectangles and leave with a shape akin to an I-beam. By the 1980s, blooming mills were being replaced by continuous bloom casters, which Bethlehem opened in Steelton, Pa., for its rail products. Steelmakers that dedicated resources to remain competitive in structural shapes invested in electric furnace (scrap-based) steelmaking shops that were married to continuous bloom casters producing beam blanks for even more efficient rolling. Not viewing structural steel as a "core" product, Bethlehem devoted its scarce capital to flat-rolled products, further eroding its competitiveness in long products such as structural shapes.

2

Taking shape

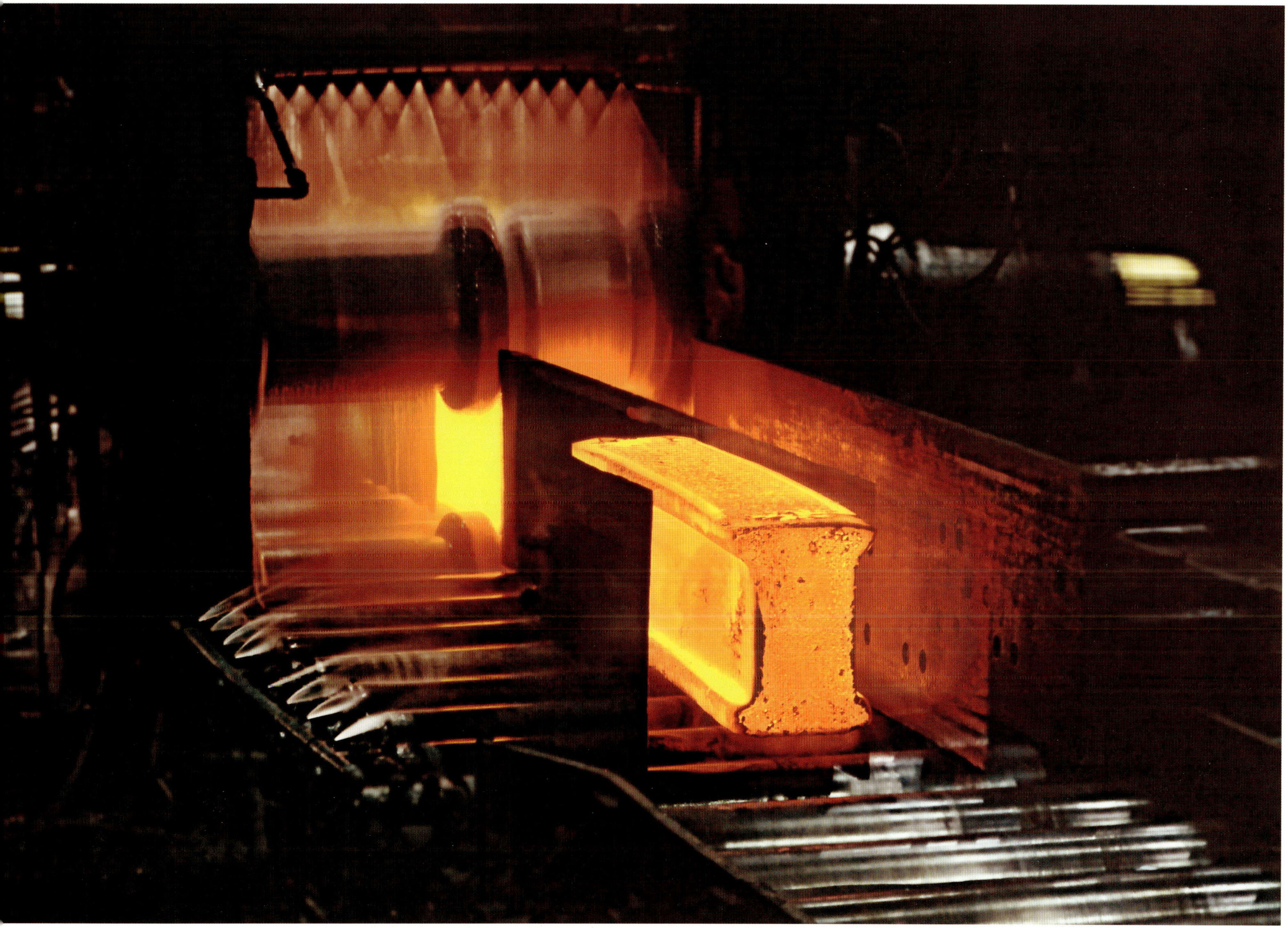

1

Looking like an I-beam

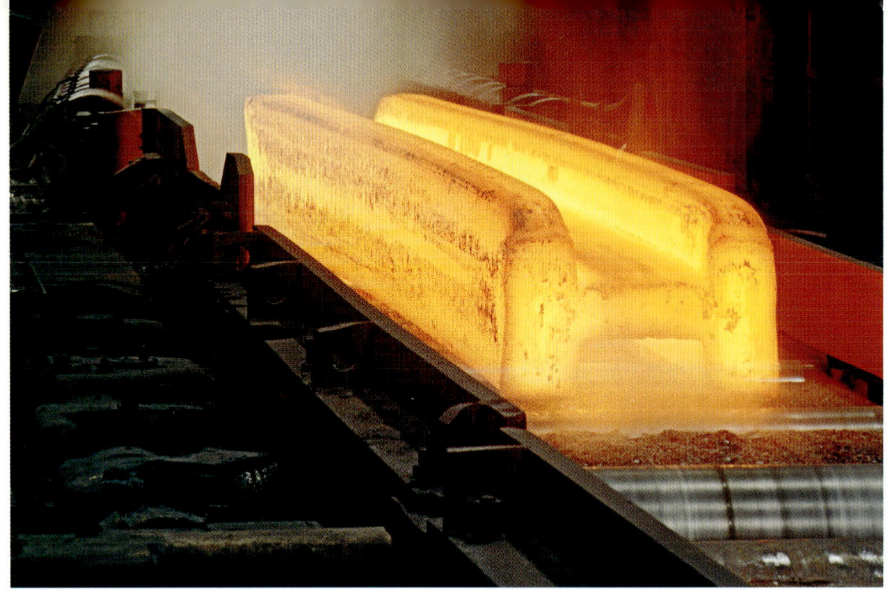

1 Once blooms entered the 48-inch mill, also known as the Grey mill, they were further reduced in width and elongated to become shapes used in building and bridge construction. The 48-inch mill operated continuously from December 1908 until October 20, 1995 and from 1910 until 1970 produced the skeletal framework for 85 percent of the New York City skyline, including such structures as the Time Life Building, the Waldorf Astoria, the Chrysler Building, Madison Square Garden but not the Empire State Building, which was the token landmark structure won by United States Steel. Ironically, one of Bethlehem's grandest signature structures, the University of Pittsburgh's Cathedral of Learning, stands in the shadow of U.S. Steel's headquarters in Pittsburgh.

2 The 48-inch mill, also known as the Grey mill, began rolling in December 1908, two years after Bethlehem Steel's founder and first chairman Charles M. Schwab gambled on technology produced by Englishman Henry Grey. Placing more than $5 million on the line, he said to his assistant, "Wardie, if we are going to go bust, we will go bust big." Schwab was always the gambler, and this wager paid off handsomely. Bethlehem Steel became the first – and largest – structural steel producer in the United States, and its products helped usher in the age of the skyscraper and the long-span bridge. During the course of the mill's lifetime, few technological changes were made. The original steam engines were still operating when the corporation invested $50 million in the mid 1980s to update and electrify the first mill stand (the roughing stand, so called because it established the rough shape of the finished product). The new mill stand brought a high level of computerization and sophistication to the mill to help roll steel to the most precise dimensions. (Note: Charles M. Schwab's initial $5 million investment in 1908 and the $50 million invested in the mid-1980s are each equivalent to about $100 million in 2007.)

2

Roller in control,
Grey Mill pulpit

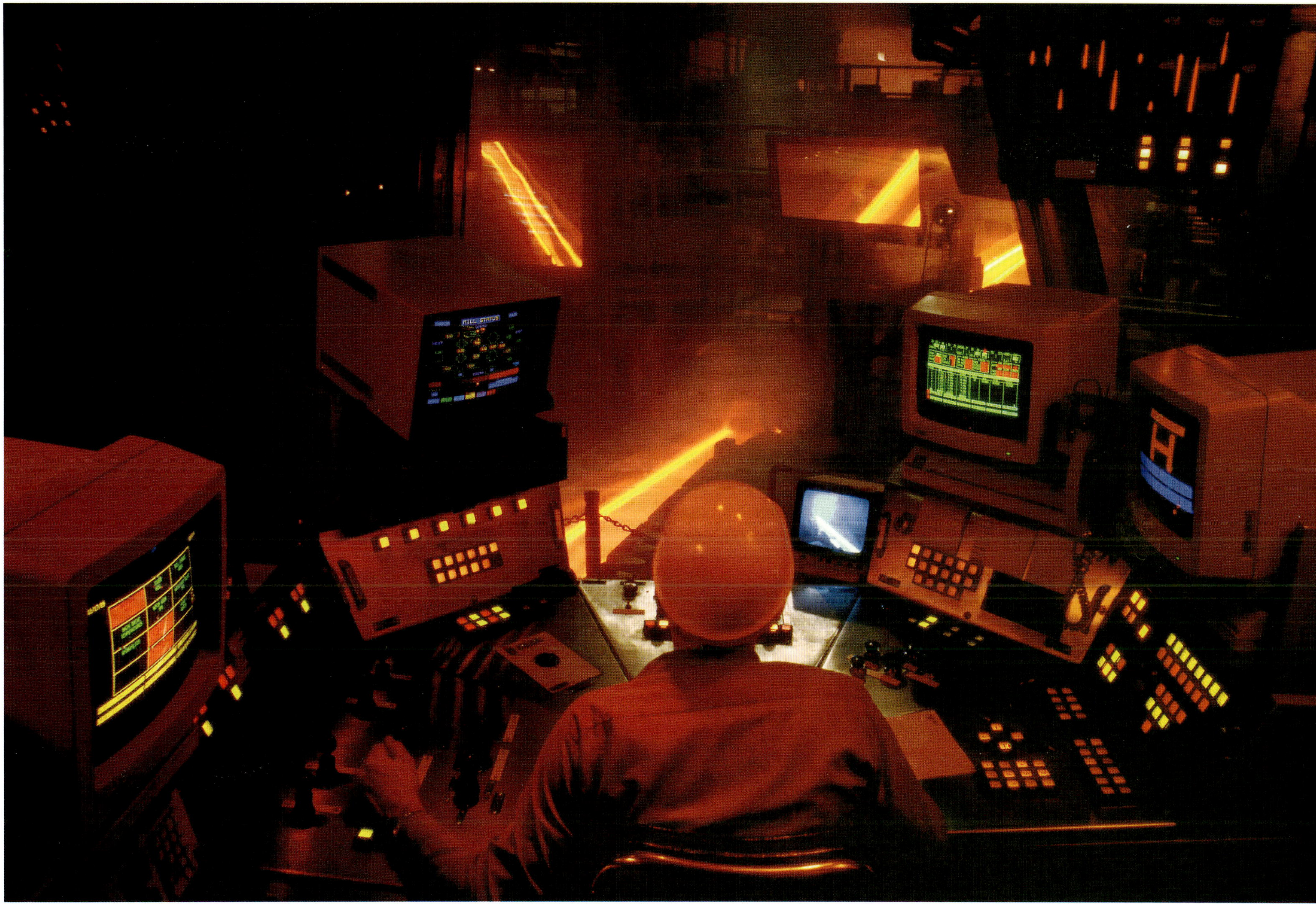
MILL STATUS

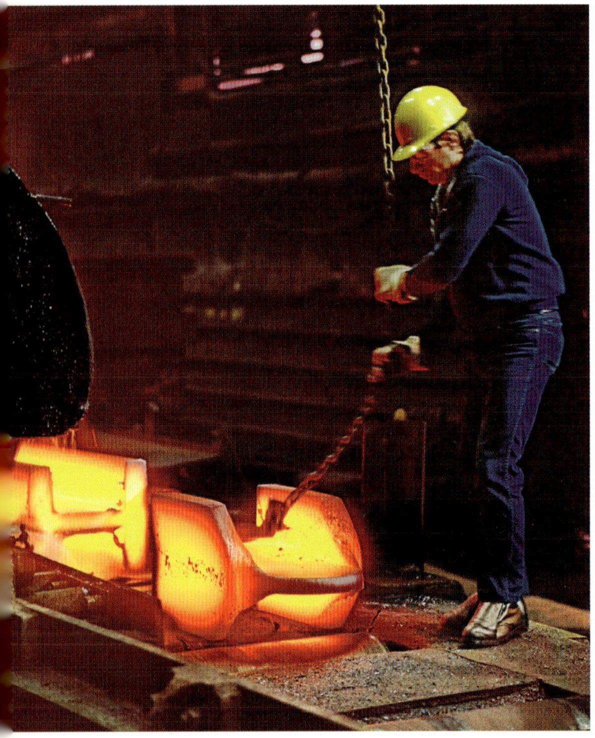

2

Hooking a jumbo beam-crop

1 Although after 1970 Bethlehem was no longer the dominant player in the worldwide marketplace for structural products, it remained a significant force, producing the broadest range of structural products long into the 1980s, including this heavy column for high-rise construction. Not long after the 48-inch mill was modernized in 1988, mill employees developed the world's largest structural steel shape. Called the meter beam, the flange was up to 40 inches deep, providing the member with the strength to handle loads of high-rise construction. Early demand for the meter or jumbo beam was strong – until two things happened in the early 1990s. First, Nucor, Bethlehem's major competitor for structural shapes, developed a similar beam, which Nucor could sell at a lower price because of its lower cost structure. Second, the market for high-rise construction dried up. Nucor captured what little demand remained for the jumbo beam. With virtually no market for the large beams rolled at the 48-inch mill, Bethlehem's board of directors elected to close the mill in the fall of 1995, concurrent with the closing of basic steelmaking at the "home" plant of the corporation.

2 After the steel was rolled into the desired shape at the Bethlehem, Pa., plant's structural mills, the ends were cropped to leave a clean edge. The ends of this wide-flange beam, along with all other ends, were then recycled back into the steelmaking process. While the basic oxygen steelmaking process used primarily molten iron, about 10 percent of the heat of steel was scrap steel, including cropped ends. This employee is wearing all the standard and appropriate personal protective equipment, including a hard hat, side-shield safety glasses, and metatarsal shoes. The shoes have steel inserts covering the toes and a steel instep to guard against crushing injuries. Although not visible, the employee was probably wearing hearing protection, little sponge ear plugs in his ears to minimize the intense clatter of the heavy sections as they moved through the mill. When personal protective equipment was mandated in the contracts between the nation's largest steelmakers and the United Steelworkers of America in the 1950s, management set up shops within their plants for employees to obtain the appropriate safety equipment.

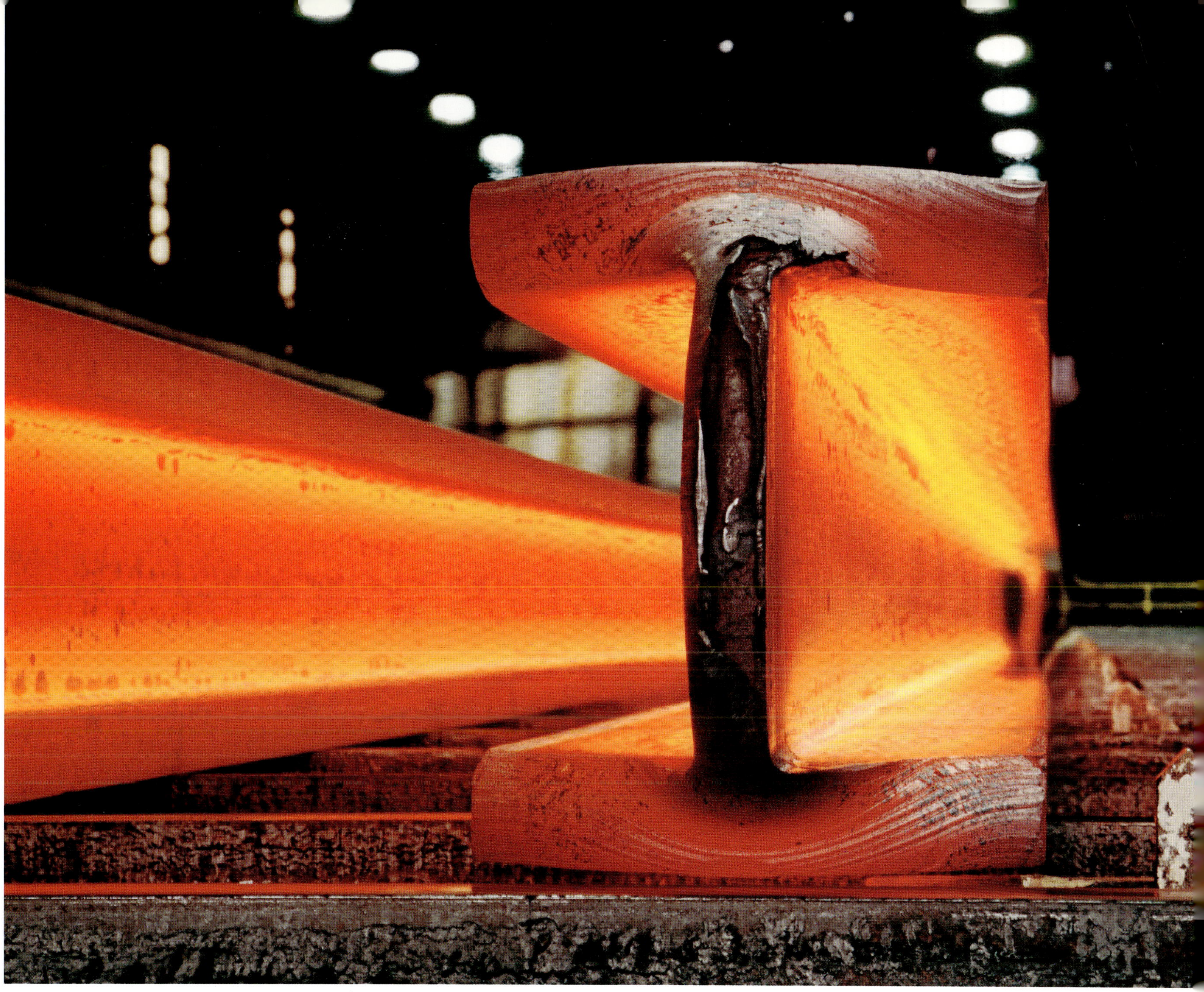

1

Wide-flange beams on the cooling bed

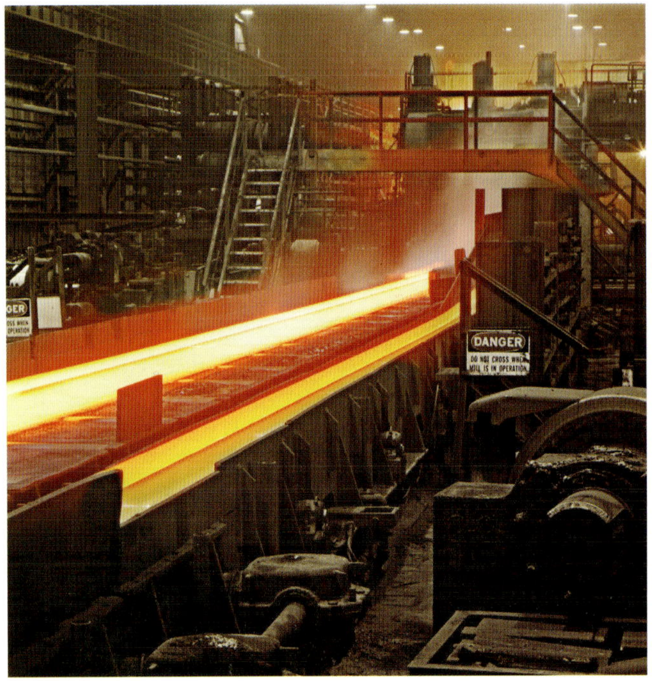

2

Rolling H-pile in the Combo Mill

1 The combination rolling mill was one of two major capital projects at the Bethlehem, Pa., plant completed in 1968 (the other was the basic oxygen furnace shop). The combo, as it was called, consolidated light and medium wide-flange production and gave the plant added flexibility in producing other sections such as channels and angles, sheet piling, h-piling, and rail-car building sections. When opened, the mill contained a high degree of technology to help roll precisely to customer specifications. A series of four mill stands took the incoming "beam blanks" and elongated and formed the steel to create the desired cross section. Mirrors on mill stands allowed operators to monitor the process in front of and behind the stand. The combination mill allowed Bethlehem to remain an active player in the structural shapes market long after other fully integrated steelmakers had called it a day. Bethlehem Steel was the last domestic steel company to produce the full range of structural shapes including piling.

2 As steelmaking was ending at the Bethlehem plant in the fall of 1995, Bethlehem Steel's board of directors authorized $26 million to improve productivity at the combination mill. A cornerstone of the modernization was sophisticated computer systems that were supposed to be installed and functional during a six-week shutdown of the mill after the closing of basic steelmaking. Employees from other operations transferred into the mill after their "home" operations were closed and were frustrated by the delays caused by the new systems. The modernization took much longer than anticipated, and the computer systems were slow in ramping up. It wasn't until the late spring of 1996 that production speed at the mill was able to reach its anticipated potential. By then, customers had defected.

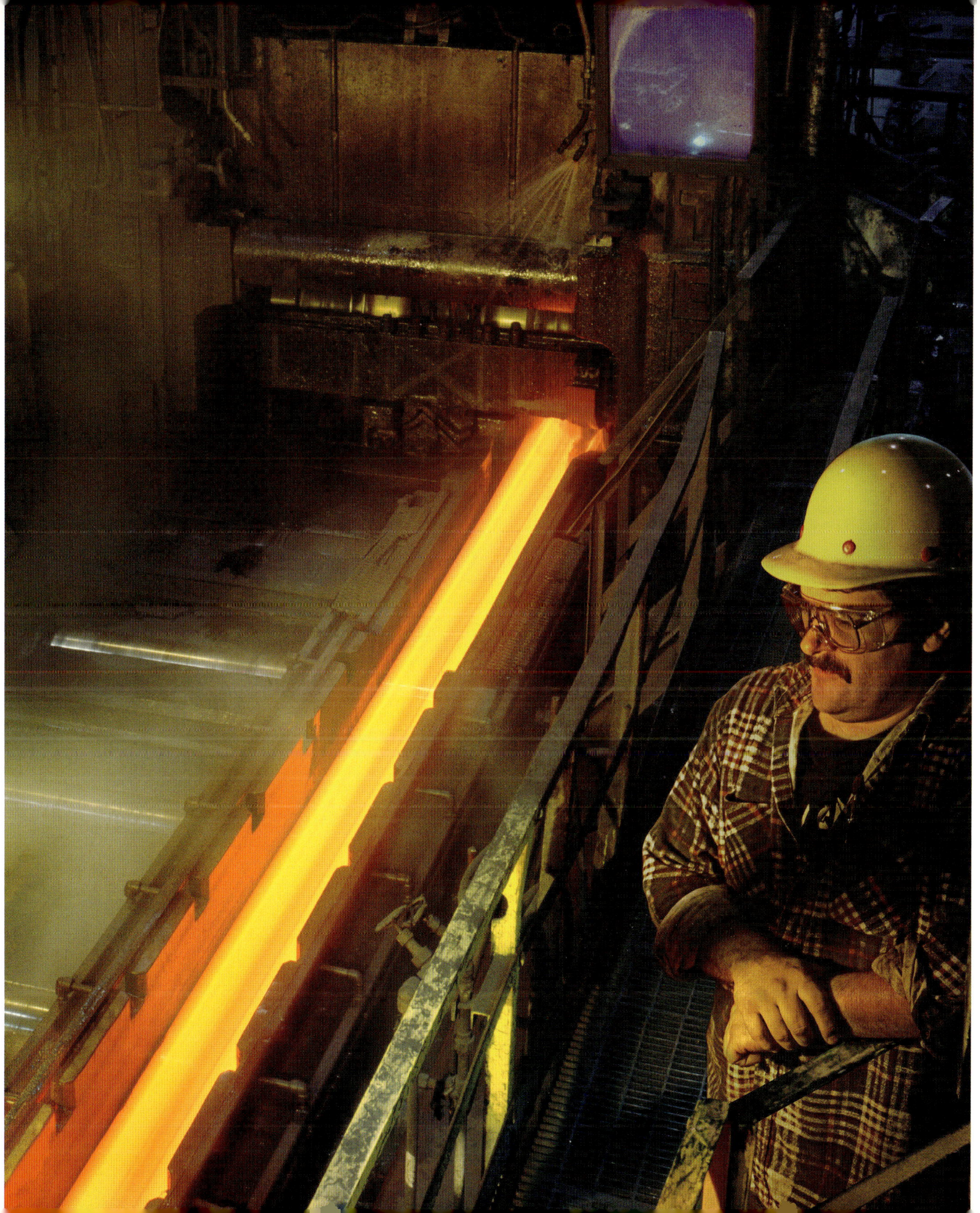

1

Keeping an eye on a Z-pile section

1

Z-pile on the cooling bed

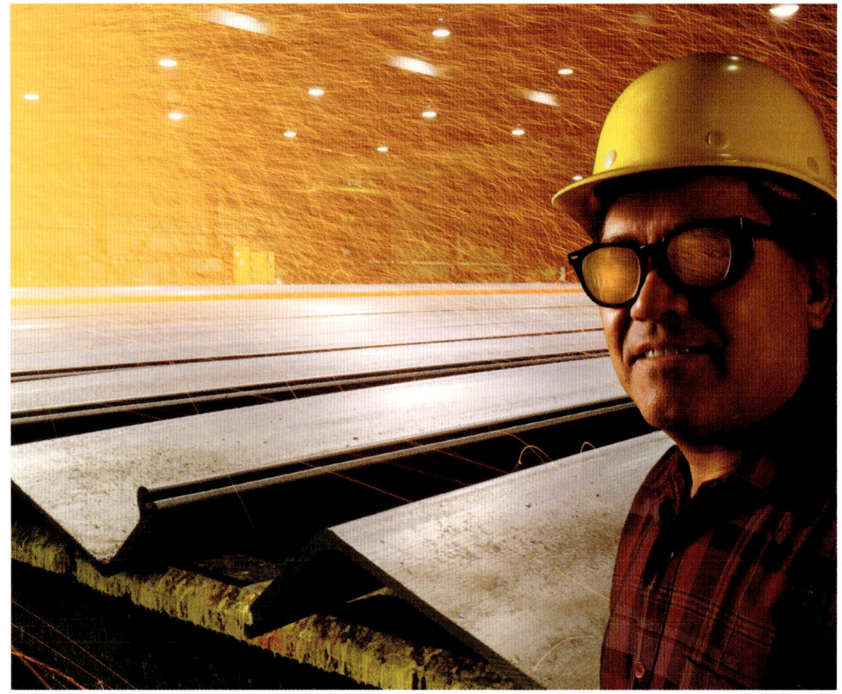

1 Rollers, such as Jose Santiago, were instrumental in the development of techniques to bring new products to the marketplace. In the 1980s and 1990s, Bethlehem Steel's combination mill expanded its offerings in piling, which is used for marine applications such as cofferdams and for sound barriers. Z piles, pictured here, have interlocking ball and socket closures, allowing for water-tight and sound-proof performance. As the mill's fortunes and future were fading, sales and marketing personnel emphasized piling products as a niche for the company – the largest product offering in North America. Newer competitors were capturing the markets for most structural products, and in January 1997 Bethlehem Steel announced the mill's March closing. The key competitive differences between Bethlehem Steel and its competitors were that newer companies had logically aligned steelmaking and casting facilities and did not shoulder health-care and pension benefits for people who left their companies through layoff or retirement.

2 It does not get any louder or more dramatic than at the hot saw where end cropping and customer sizing takes place. Bethlehem had close relationships with fabricators that provided a variety of services to expedite construction, such as cut-to-length and cambering. Increasingly during the late 1980s and throughout the 1990s, Bethlehem adopted a just-in-time delivery philosophy to ship material only when the fabricator was ready to receive it. That meant a much tighter schedule, including shipment as soon as possible after the rolling was complete, which minimized unpaid inventory.

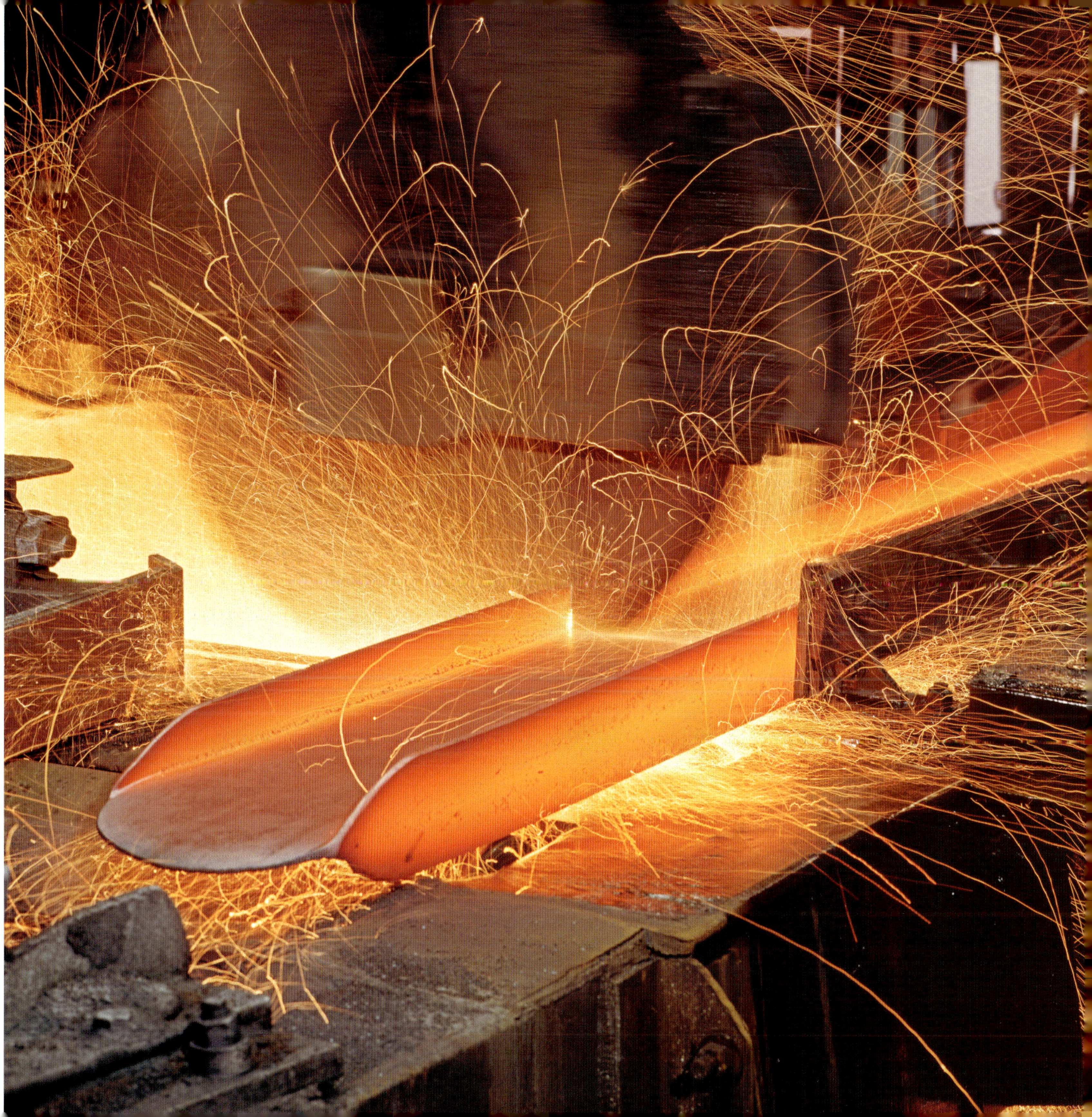

2

The hot saw
cropping an I-beam

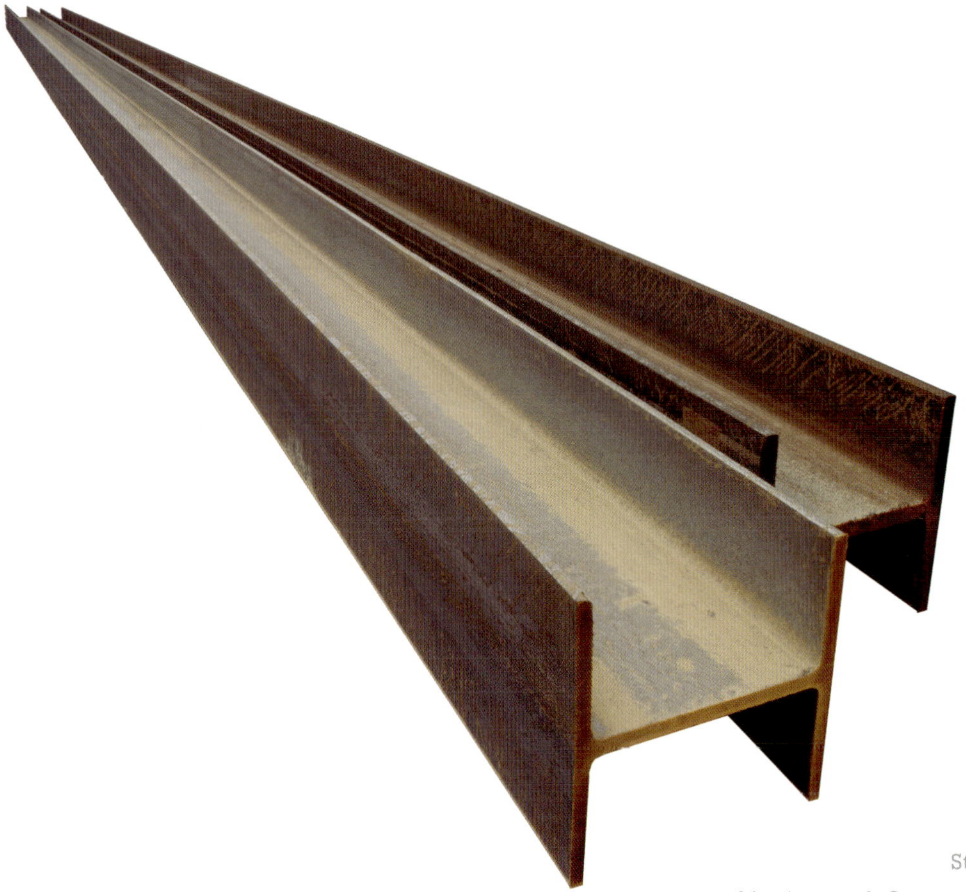

Structural products from all rolling mills at the Bethlehem, Pa., plant were shipped from a 29-acre shipping yard. Once exiting the 48-inch or combination mill, the steel was stored for quick retrieval and shipment by rail or truck to customers. Overhead crane runways spanning the yard, moved steel overhead while straddle carriers, large vehicles that could lift and transport beams in their center sections, moved steel on the yard's roadways. A sign at the entrance to a pedestrian tunnel through which most beam yard employees walked read, "Entrance to the Largest – Best – Structural Shipping Yard in the Free World – Let's keep it that way!" The sign was erected in the 1960s and outlived the structural mills and shipping yard.

Beam yard loading

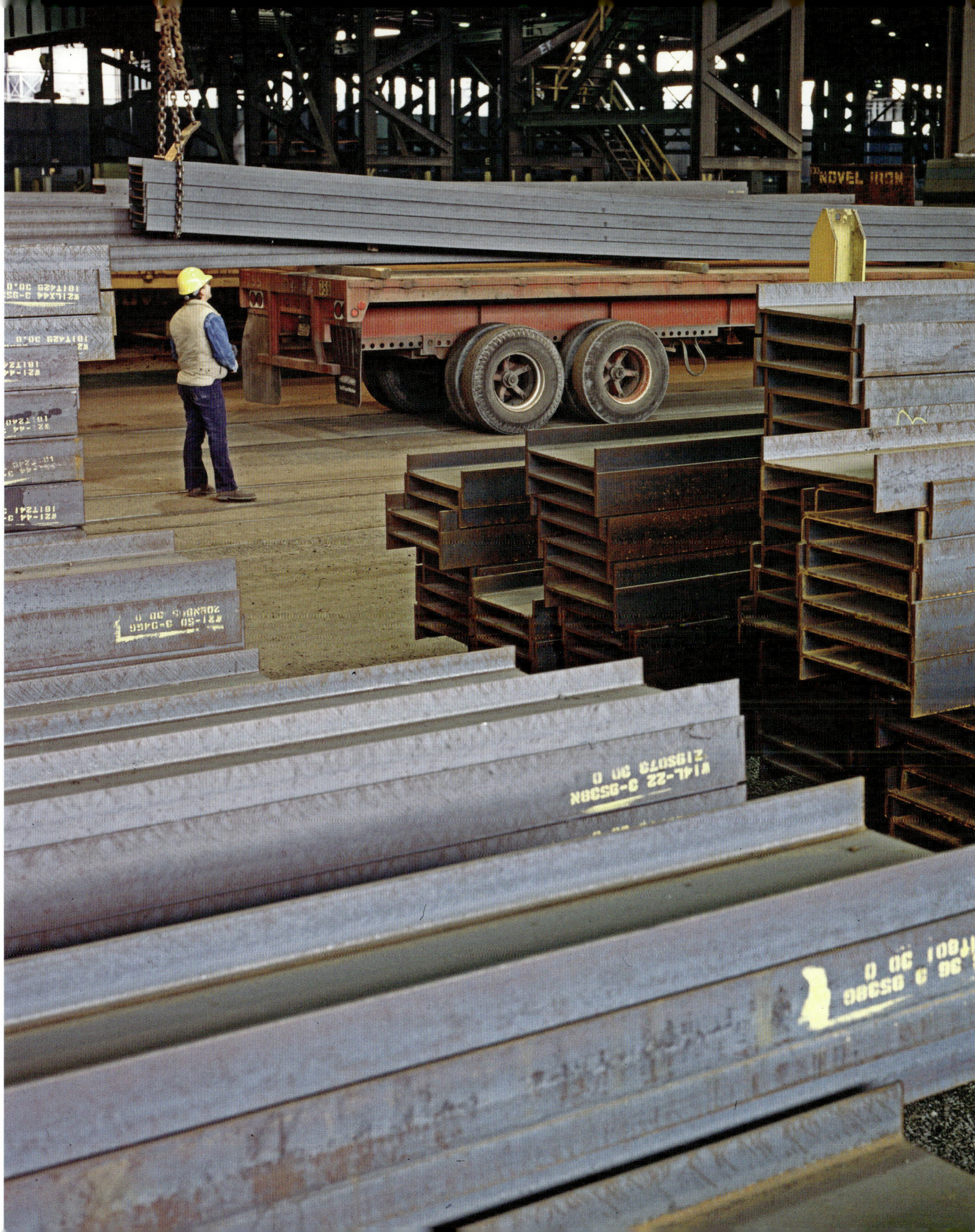

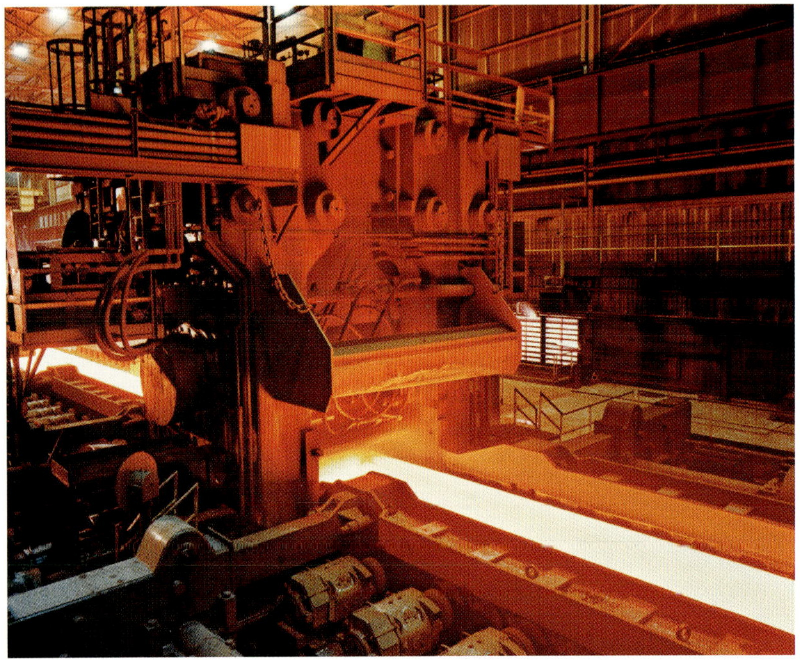

A long plate in the 110" Plate Mill

Plate products are a staple in buildings, machinery, ships, and barges. The 160-inch plate mill was one of the first facilities in operation when Bethlehem Steel opened its Burns Harbor, Ind., plant in 1964. Once basic steelmaking facilities were added to the plant in the early 1970s, Burns Harbor quickly became the jewel in Bethlehem Steel's crown. Bethlehem added the 110-inch plate mill, pictured here, in the 1970s to widen its plate product line. The flat-rolled products – sheets for automobiles and appliances and plates for construction equipment – produced at Burns Harbor became the strength of Bethlehem Steel as it closed facilities for long products in the 1980s and 1990s. Bethlehem further strengthened its position in plate products by acquiring Lukens Steel in 1998, making Bethlehem the largest and most diverse plate producer in the United States.

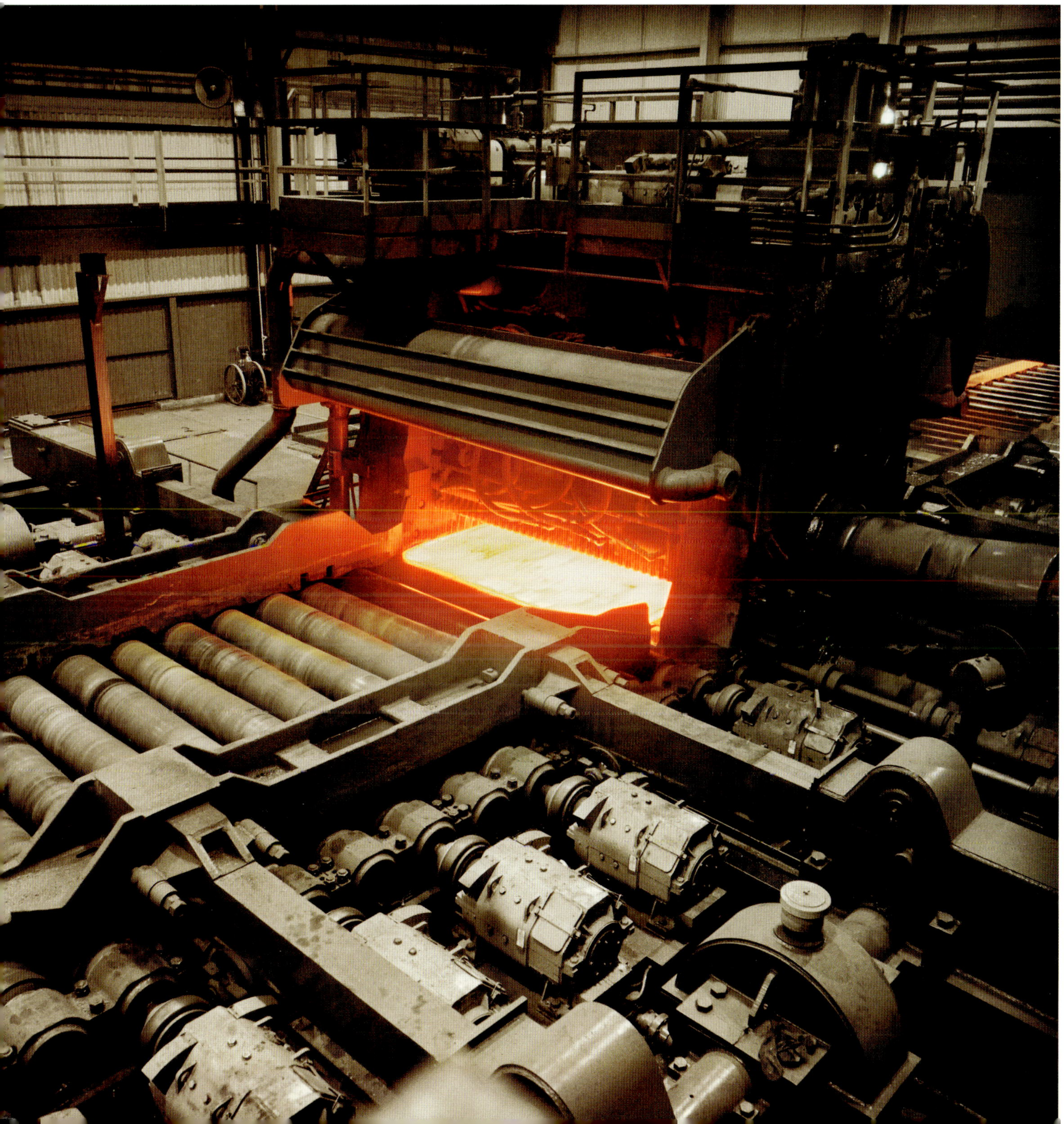

A new slab enters the 110" Plate Mill

1

A very dangerous situation

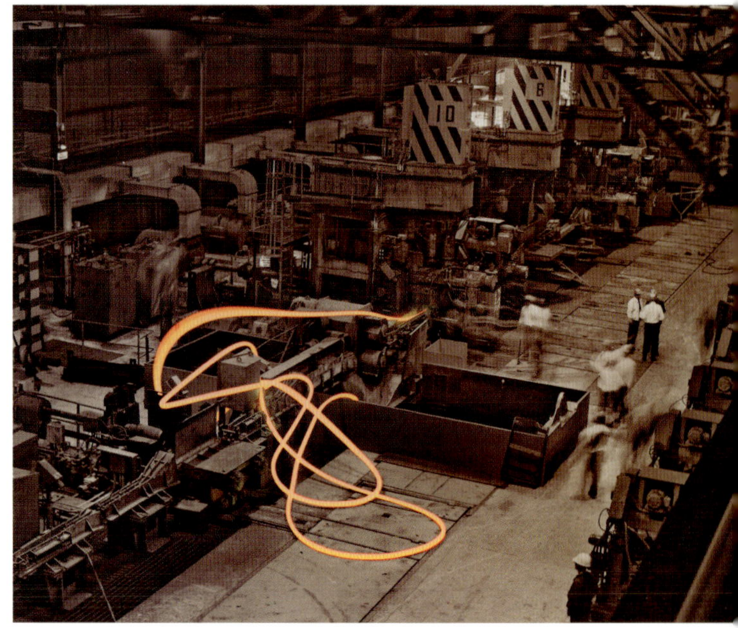

1 This "cobble" or a piece of rolled steel that had escaped the rolling line, while moving through the mill at nearly 60 miles per hour, occurred during startup and ended up as scrap.In the late 1970s, Bethlehem invested in a new 13-inch bar mill at its Lackawanna, N.Y., plant along Lake Erie to serve the automotive industry in Detroit and elsewhere in the midwest. The mill's early days were plagued with delays and false starts, in part because of widespread use of computerization not common in other finishing facilities of that era. But the mill and its people persevered, although its life as a Bethlehem facility was short lived. The corporation closed basic steelmaking and finishing operations in October 1983 and then supplied the mill with billets produced at the Johnstown, Pa., plant. The 200 miles between the Johnstown and Lackawanna plants added to Bethlehem's lack of competitiveness just when the automotive industry, a significant customer of rolled steel, was under extreme pressure to lower costs and raise quality.

2 When it finally opened in 1981, Bethlehem's 13-inch bar mill at Lackawanna, N.Y., used very sophisticated technology. Widespread computerization allowed for a higher level of automated process control and handling than was previously available. At the end of the series of mill stands, the bars were wound onto mandrels (cylindrical shafts). The coils were off loaded from the mandrels and then automatically moved to an automated coil storage and retrieval facility, which was nine stories high. Under Bethlehem's ownership, the mill, while lauded by customers for quality and service, labored under high operating costs relative to its competitors.

3 Bethlehem Steel had a heralded history of producing "long products" used in a variety of industries. The agriculture and construction industries used reinforcing bars, known as rebar, to give shape to the other materials surrounding their products. For fast delivery to customers, Bethlehem had a huge network of reinforcing bar plants in the midwestern and eastern United States.

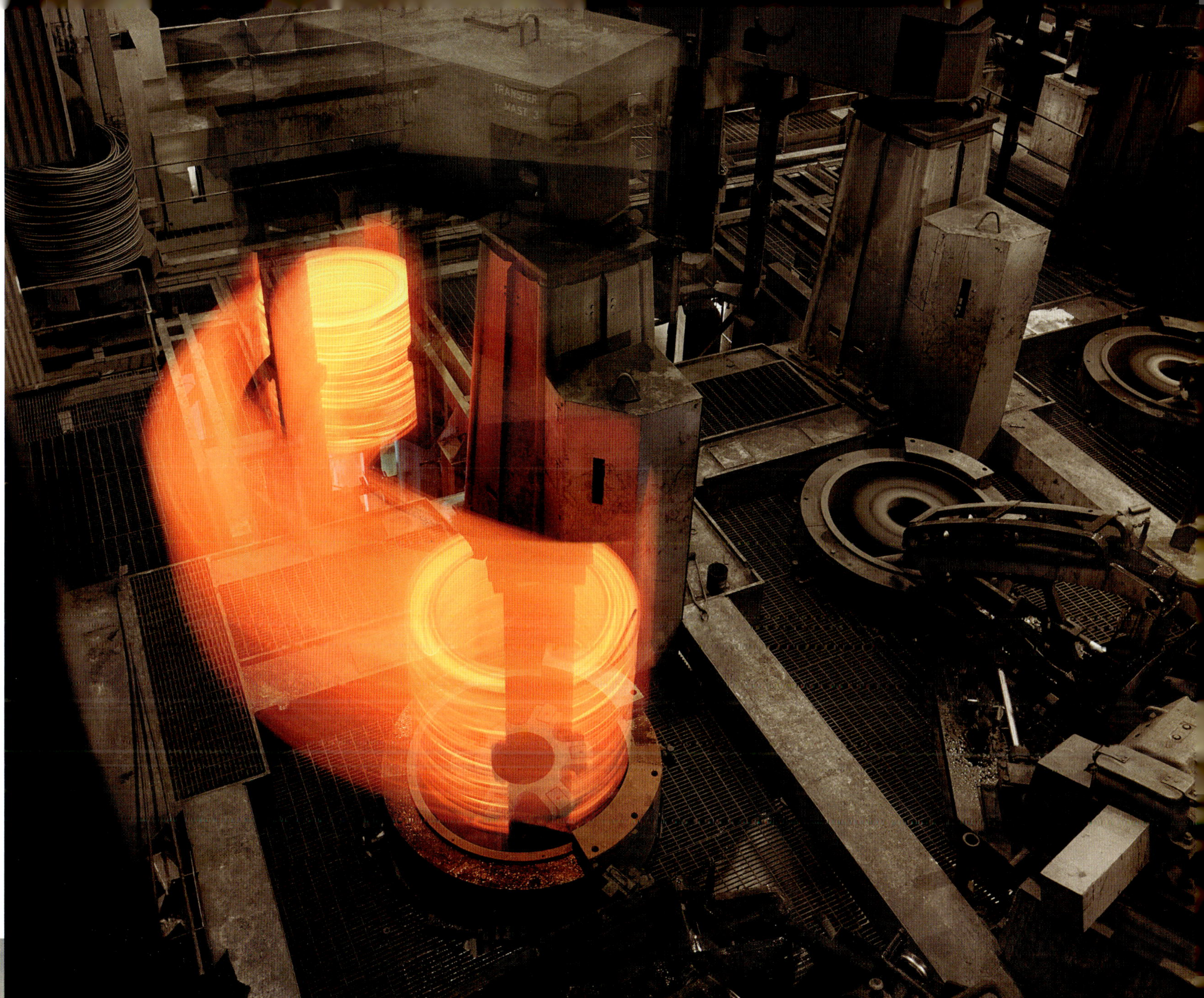

2

Robots in action

Holding a bridge together

3

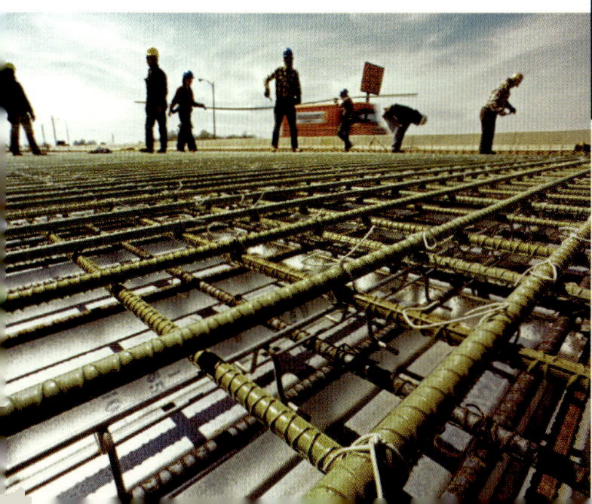

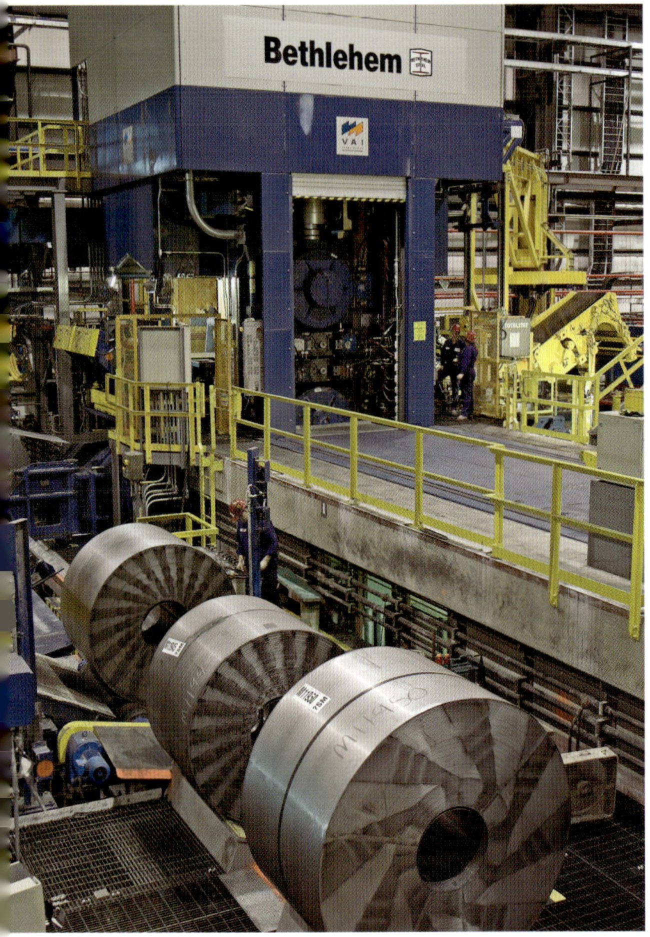

1

Maybe a new refrigerator in the Sparrows Point Cold Mill

1 For years, Bethlehem Steel's Sparrows Point, Md., plant struggled with cost and production issues for its light-flat-rolled products used for metal building, appliance, and automotive applications. Keeping with the corporate strategy to focus on flat-rolled products, in 2000 Bethlehem invested more than $325 million to open a new cold finishing mill at Sparrows Point. Incoming steel from the plant's hot strip mill was fed into the five-stand tandem mill running for 1,025 feet. At the end of the line, the steel was at its intended gauge and wound into coils for further processing. Although Bethlehem's logo changed very little during its 99-year life, it was not displayed frequently inside facilities. When the assets of Bethlehem Steel were sold to International Steel Group in May 2003, in less than one week the new owner removed all traces of Bethlehem's signage, including the one pictured here.

2 In the 1980s, Bethlehem Steel installed process control systems, as money permitted, to meet the increasing demands of customers, particularly those for light-flat-rolled sheets. The coating lines at Sparrows Point, Md., were updated extensively in the early 1990s to meet the needs for greater uniformity and diversity of coatings. Pulpit operators monitored key production parameters to ensure quality products. Extensive employee training in statistical tracking and evaluation helped improve product performance and customer satisfaction.

1

Bethlehem's last new mill starts up

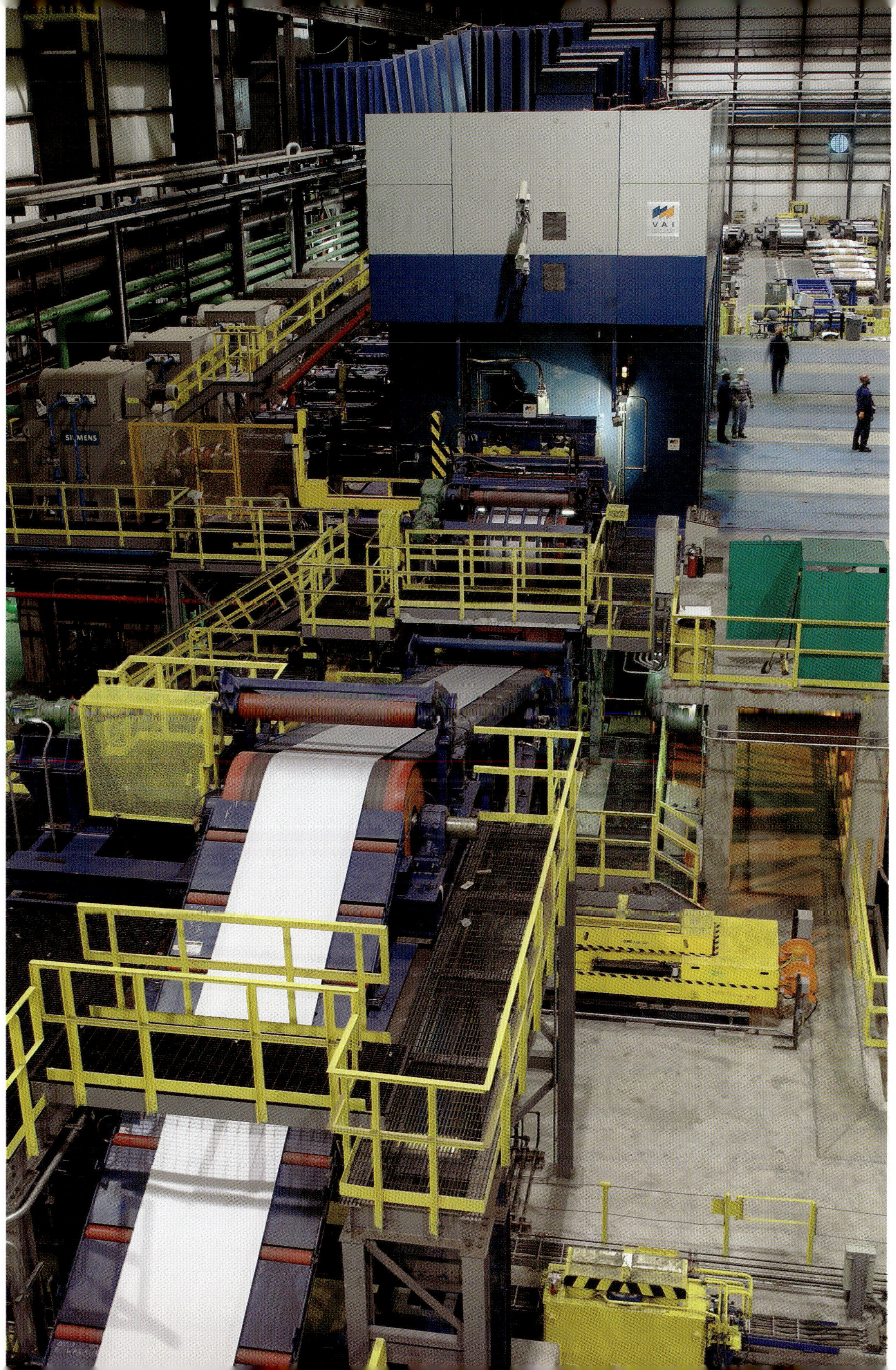

Technology invades the Sparrows Point plant

2

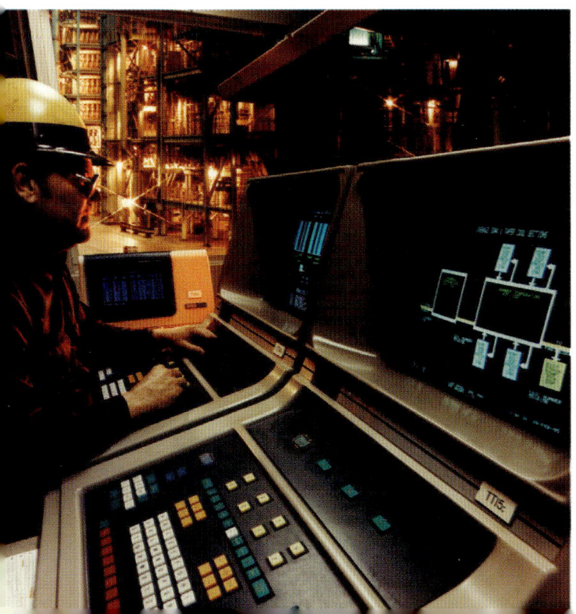

1

Acres of new steel

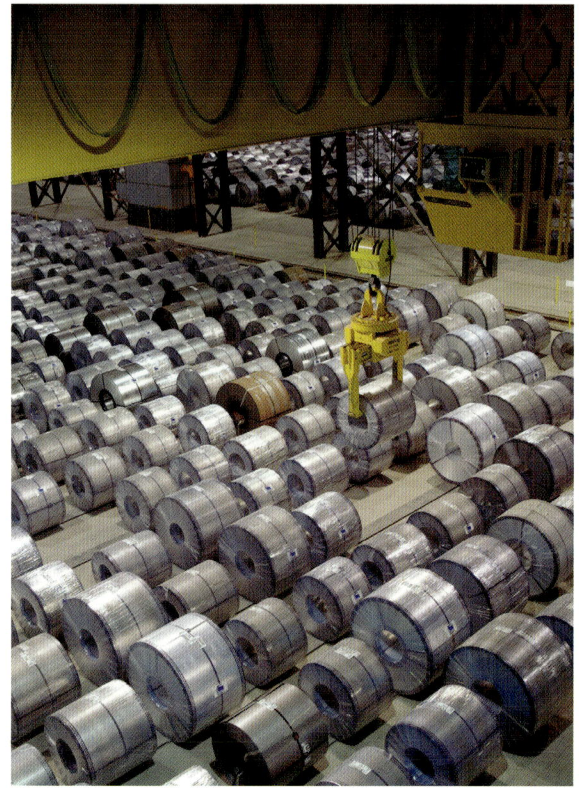

1 Coils were stored temporarily in the middle of the mill until they were either transferred to one of Sparrows Point's coating lines or further processed elsewhere in the cold mill. Nineteen overhead cranes traveled through key areas of the 800,000 square foot mill. The mill was capable of producing 1.5 million tons of steel annually. While the mill was under construction, Bethlehem was in the midst of a $600-million improvement program at Sparrows Point, Md., which included relining the giant "L" blast furnace and modifying the basic oxygen furnace and slab caster. At the peak of all construction projects during the summer of 1999, the number of people working on the plant's property doubled from 4,000 (Sparrows Point employees) to 8,000 (including construction contractors).

2 Once ready for delivery to customers, the coils were banded and wrapped before being moved by a fully automated coil transport system. The clear recyclable packaging material around the inner and outer layers prevented rust. Customers could also specify other techniques, such as inner and outer-diameter protectors, eye bands, and belly bands, to ensure damage-free delivery. Trucks drove through the shipping area, thus alleviating the ever-present traffic bottlenecks at the older cold mill at Sparrows Point, Md.

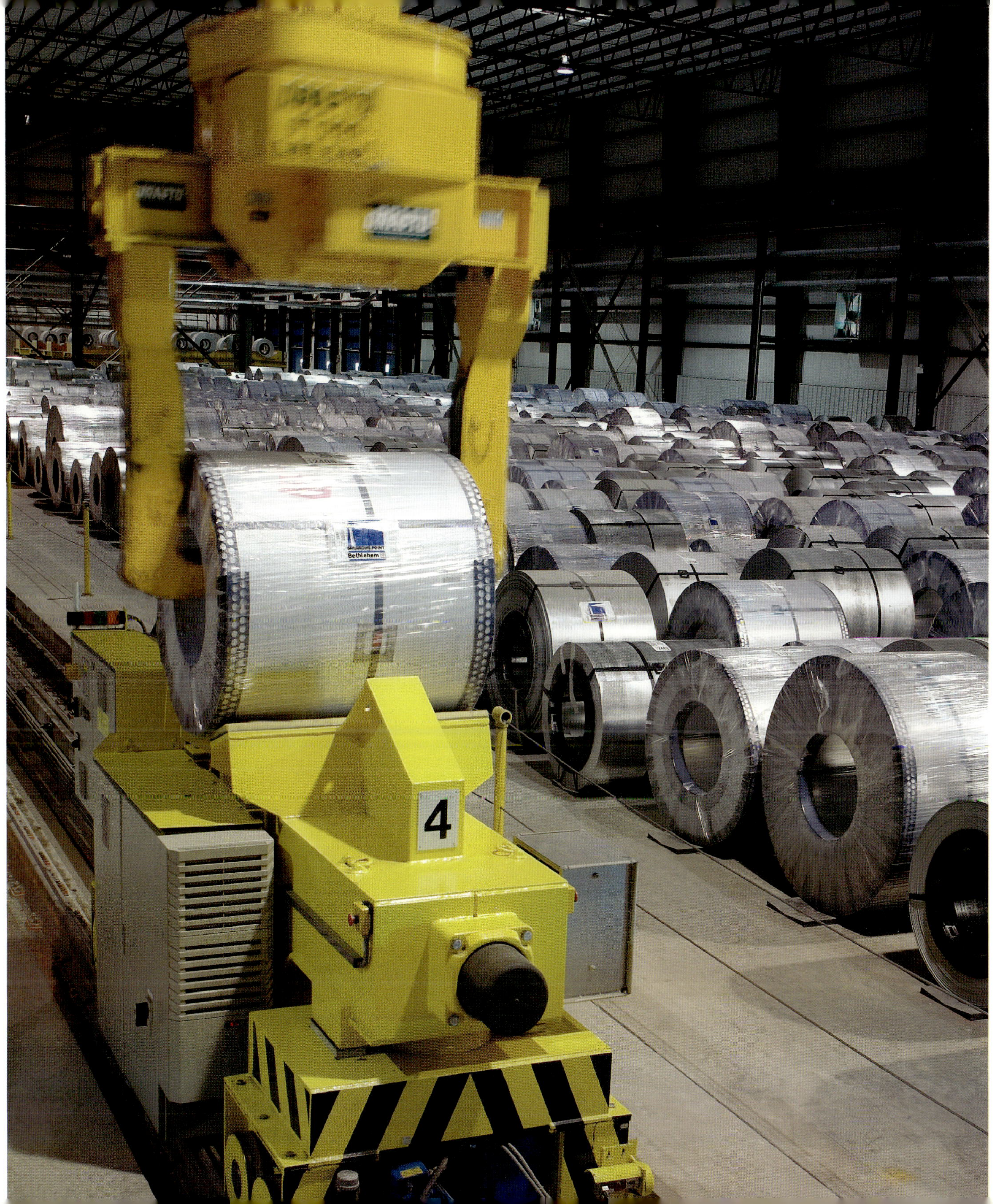

2

A robot moving coils

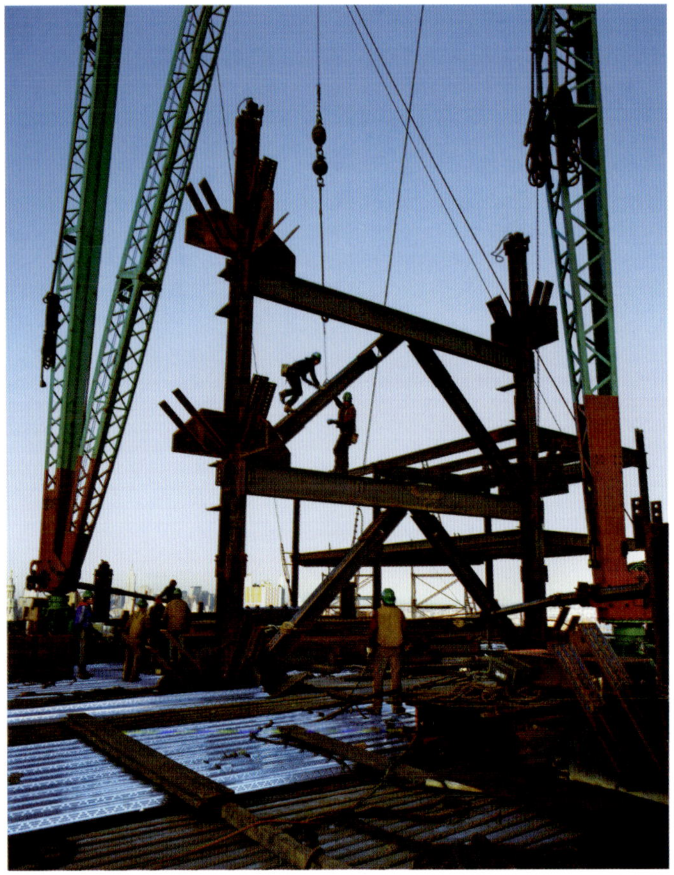

High steel over Manhattan

Bethlehem's structural products continued to factor into high-rise construction in New York City and other key areas. Citicorp's headquarters building, which featured a striking and unusual design, was built in midtown Manhattan. One of Bethlehem's largest customers for structural steel was High Steel of Lancaster, Pa. High Steel fabricated Bethlehem's wide-flange shapes for the Continental Bank building in lower Manhattan, just north of the South Street Seaport. High Steel also worked with Bethlehem's plate products for bridge construction. Bethlehem's sales training was acknowledged as the best in the industry. Sales professionals worked closely with customers and the plant employees to specify the correct steels for the end use.

The Citicorp Building, a steel masterpiece

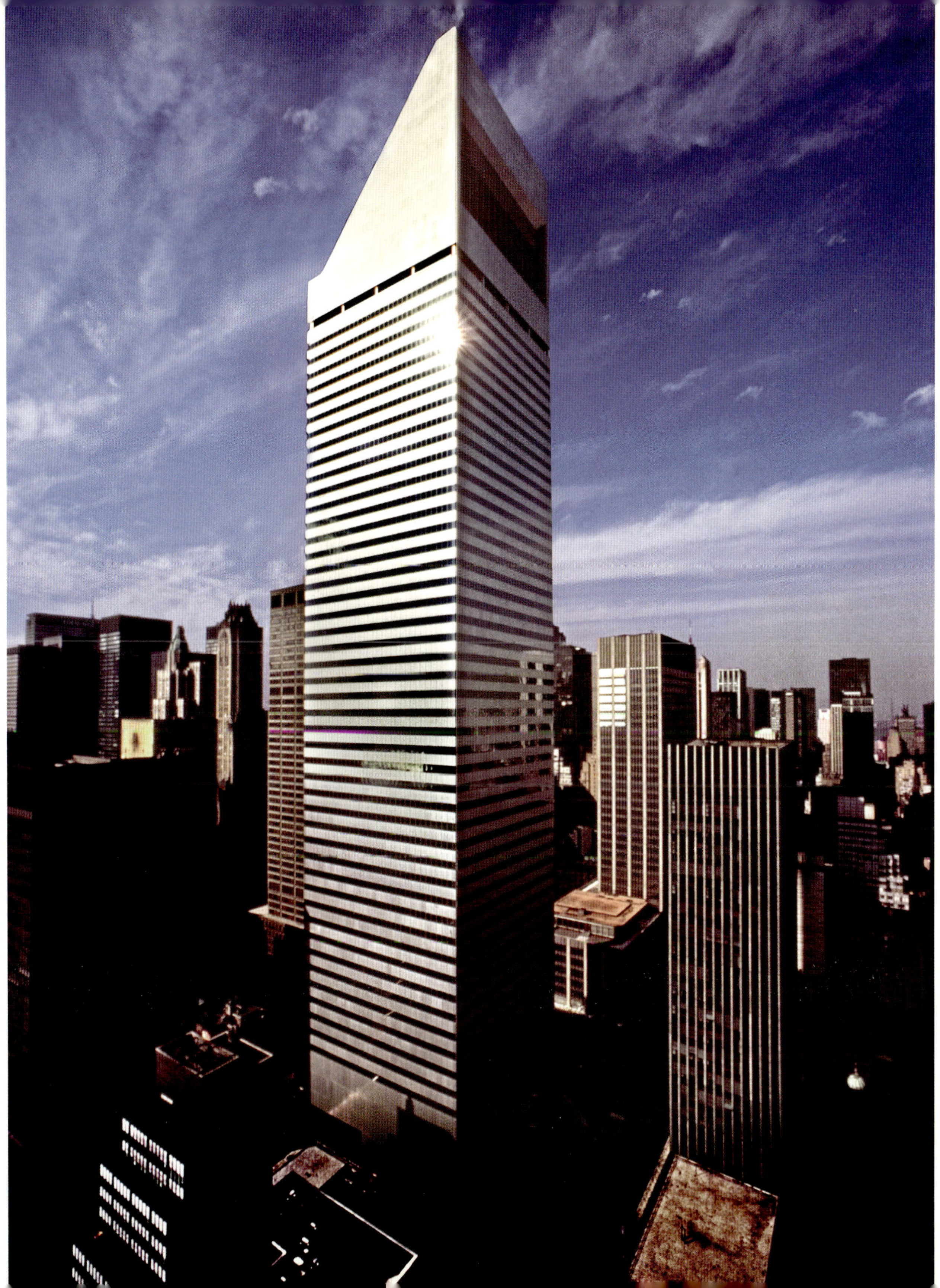

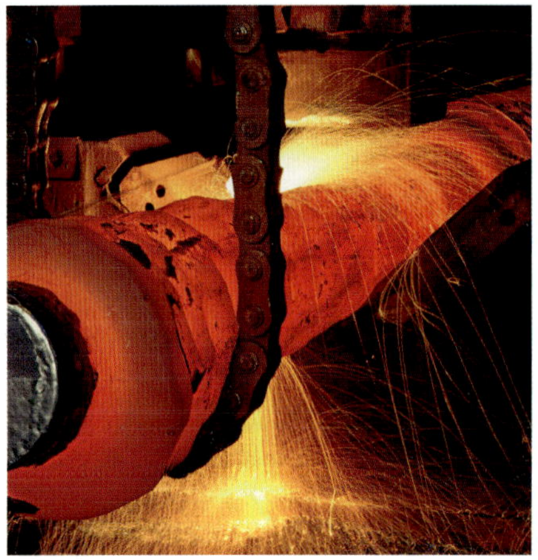

FORGING

2

Giant tongs for giant ingots

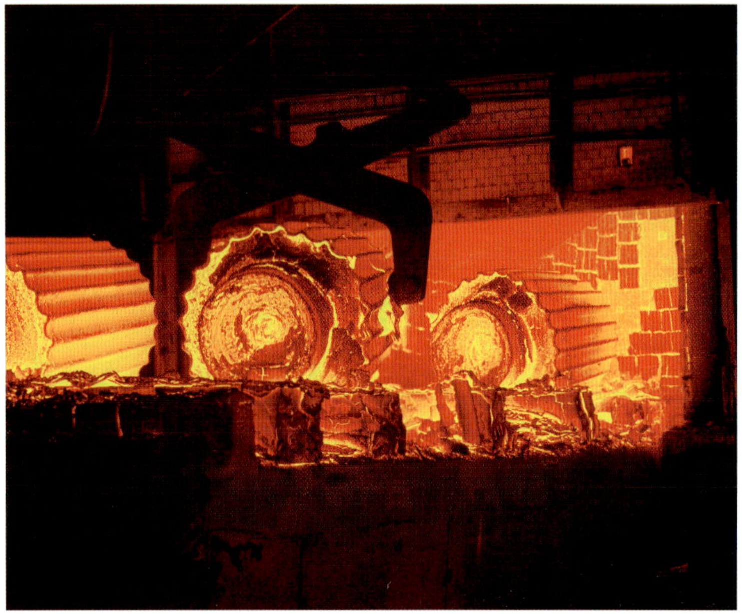

1

Flame cutting with oxygen

1 The process of forging – pounding or squeezing hot steel into shape – has changed little over the centuries. In the 1880s, the Bethlehem plant became the first manufacturer of heavy forgings in the United States. When its machine shop opened in 1887, spanning more than one third of a mile, it was the largest machine shop in the world. A massive capital project between 1885 and 1893 doubled the forging capacity at the plant. Over the years, the forging operations at the Bethlehem plant installed technology to handle increasingly larger ingots, thus greatly expanding its product line. Ingots were produced in the electric furnace shop and then reheated before forging. While the force of the forge produced the general shape of the steel, this employee was cutting away unneeded parts of the forging – known as the sync head – at the top of the ingot while the steel was still in the press. Employees working near hot metal wore flame-retardant protection, commonly referred to as "silver suits."

2 After the ingots are removed from their molds, they were reheated to the proper temperature for forging, thereby making them pliable under the forge's pressure. Ingots were removed with giant tongs and then carried through the shop to the forge, accompanied by a loud siren. The heat from traveling ingots could be felt up to 100 feet away. Like their cousins – the soaking pits used for ingots destined to become blooms – these reheat furnaces were lined with refractory brick, which needed to be replaced every few years. The brick retained the heat to reduce energy consumption and provide a fairly constant temperature in the ovens.

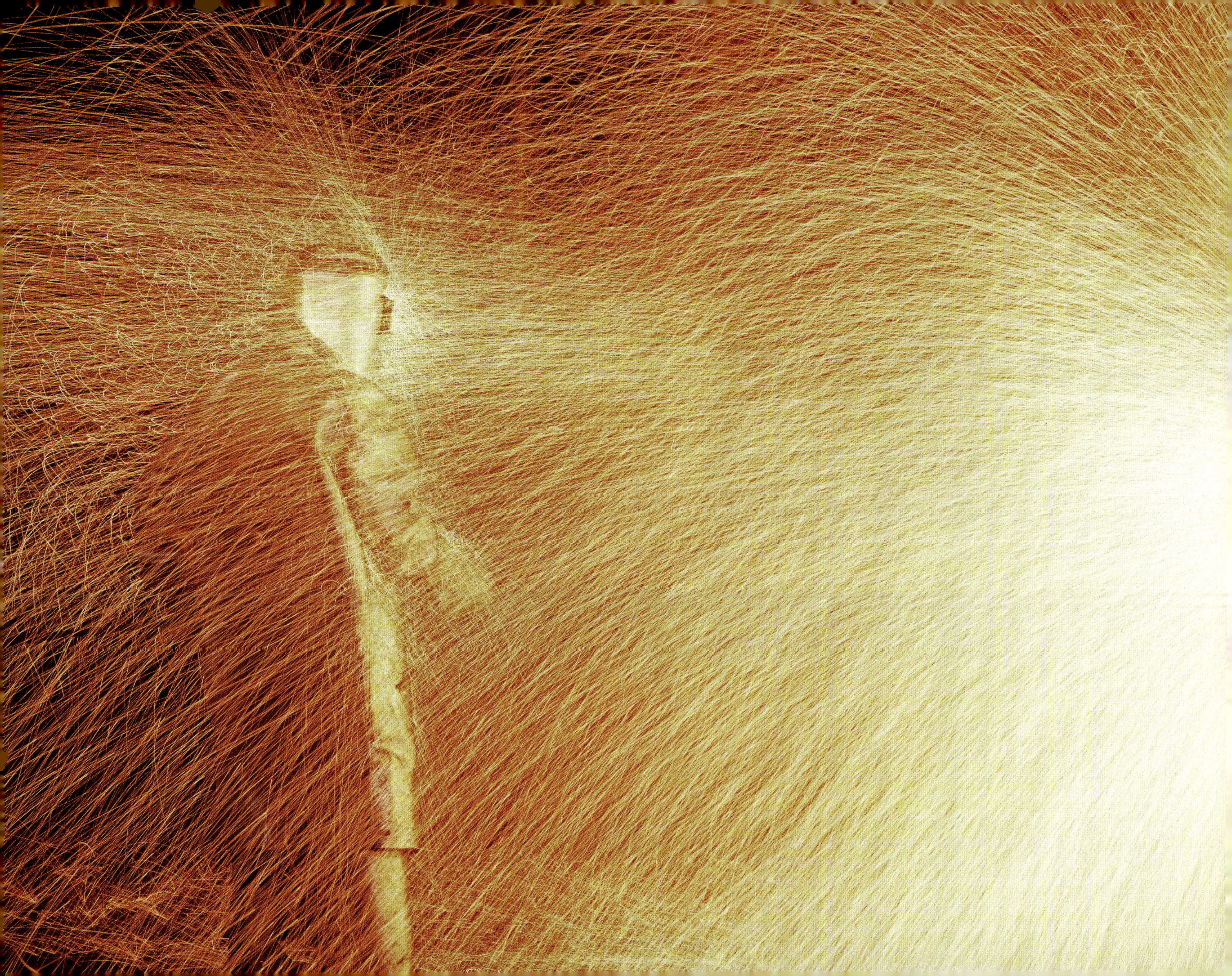

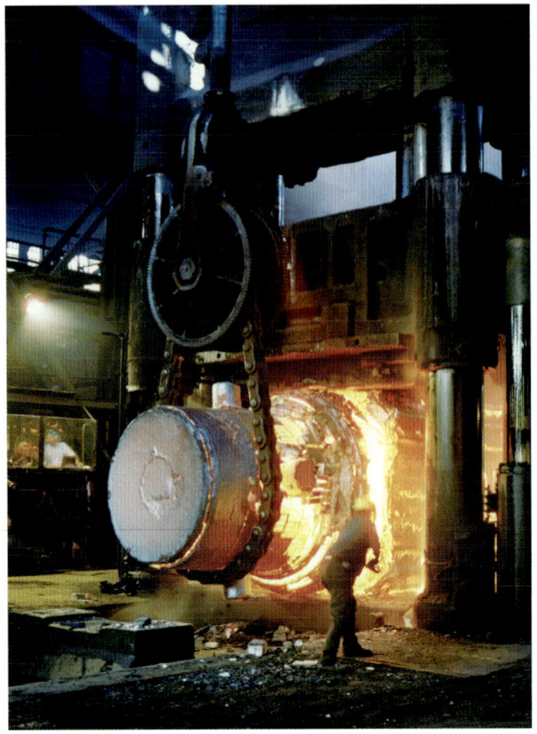

880,000 pounds of hot metal succumbing to the press

The signature workhorse of the forging division in Bethlehem was the huge No. 2 press, which provided the rough shape for the mid-to-heavier range of products. In the early 1980s, the forge was damaged, and a multi-million dollar investment was made to completely overhaul the forge. When the renovations were completed in 1984, the Bethlehem division boasted the largest open-die press forge in the western hemisphere. With 10,000 tons of force, it was capable of handling ingots up to 130 inches in diameter. Bethlehem heavily marketed the increased size and production capabilities of the press to customers, particularly those requiring large forged products common in power generation and submarines. During forging the ingots were turned and pressed continuously to produce uniform shapes.

A heated discussion

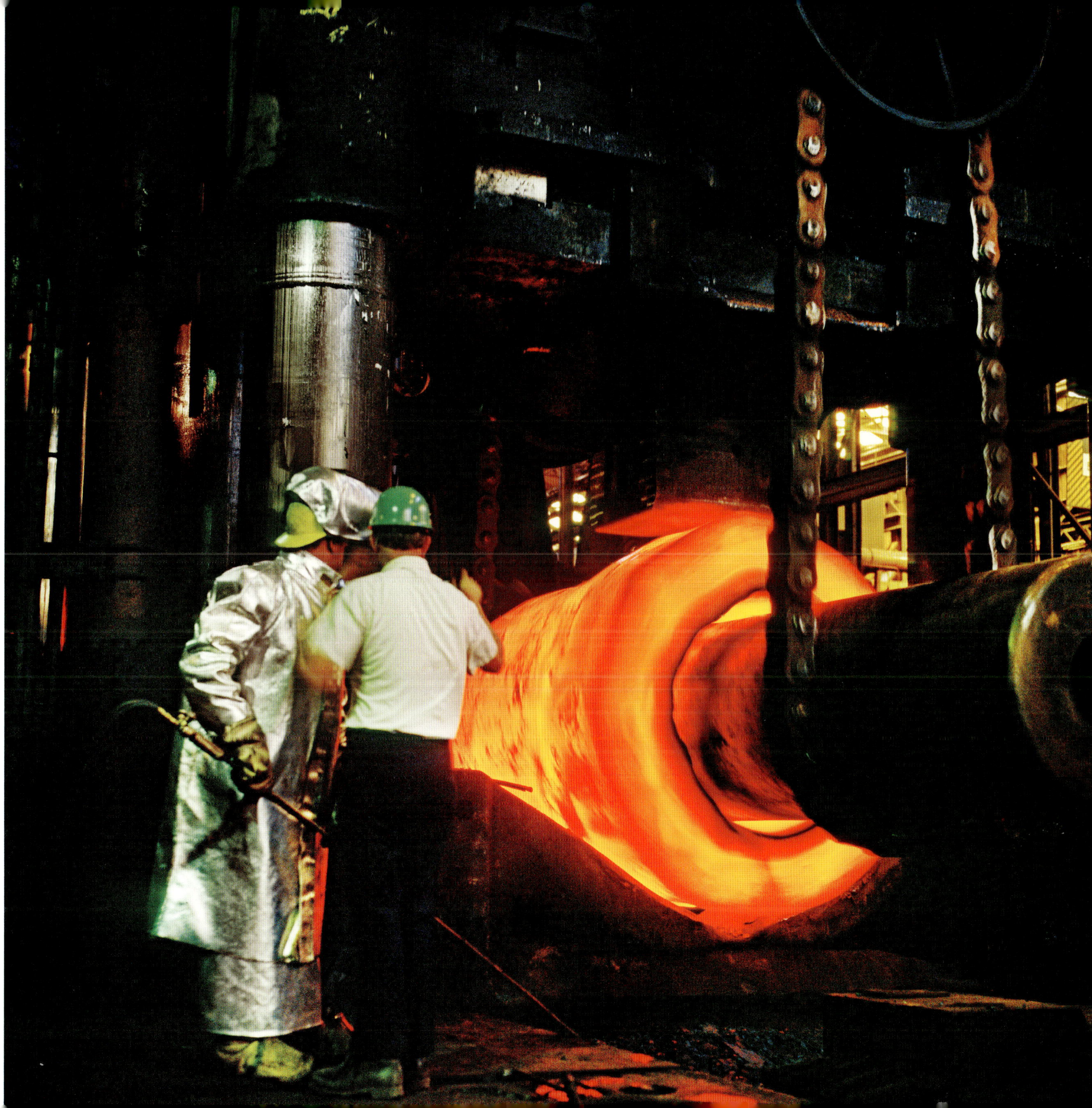

Sync head removal,
Press Forge No. 2

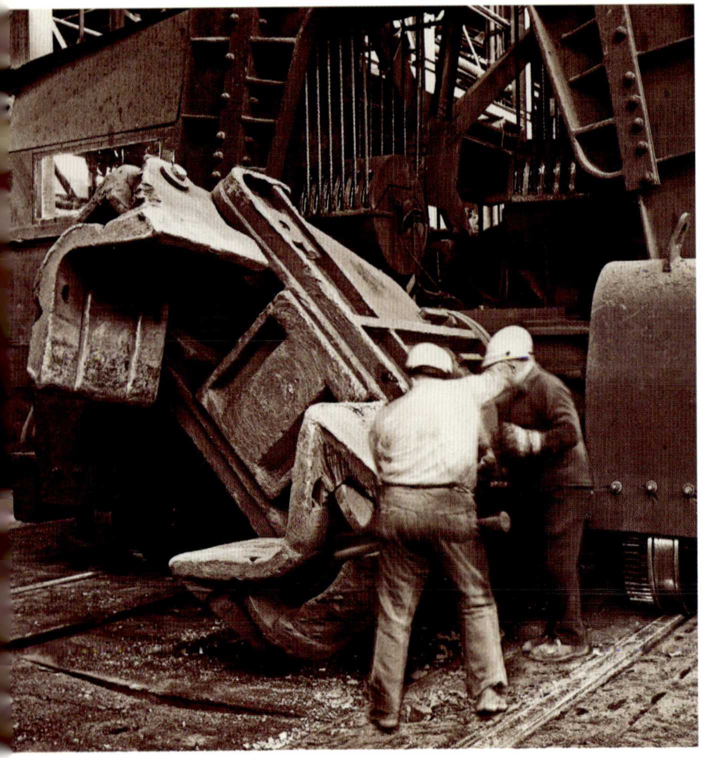

Wrestling with the jaws

These two photos were taken to document work being done for high-profile energy systems providers such as General Electric and Westinghouse. Both were large, long-time customers of forged products for turbines and other critical rotating equipment for power generation. Because of the important role played by forged products in the energy industry, customers required extensive documentation of all stages of production. One photo shows two men manually changing the manipulator at the back of the forge. The manipulator helped turn the ingot during forging. Once the mill was modernized in the early 1980s, the manipulator could be automatically changed.

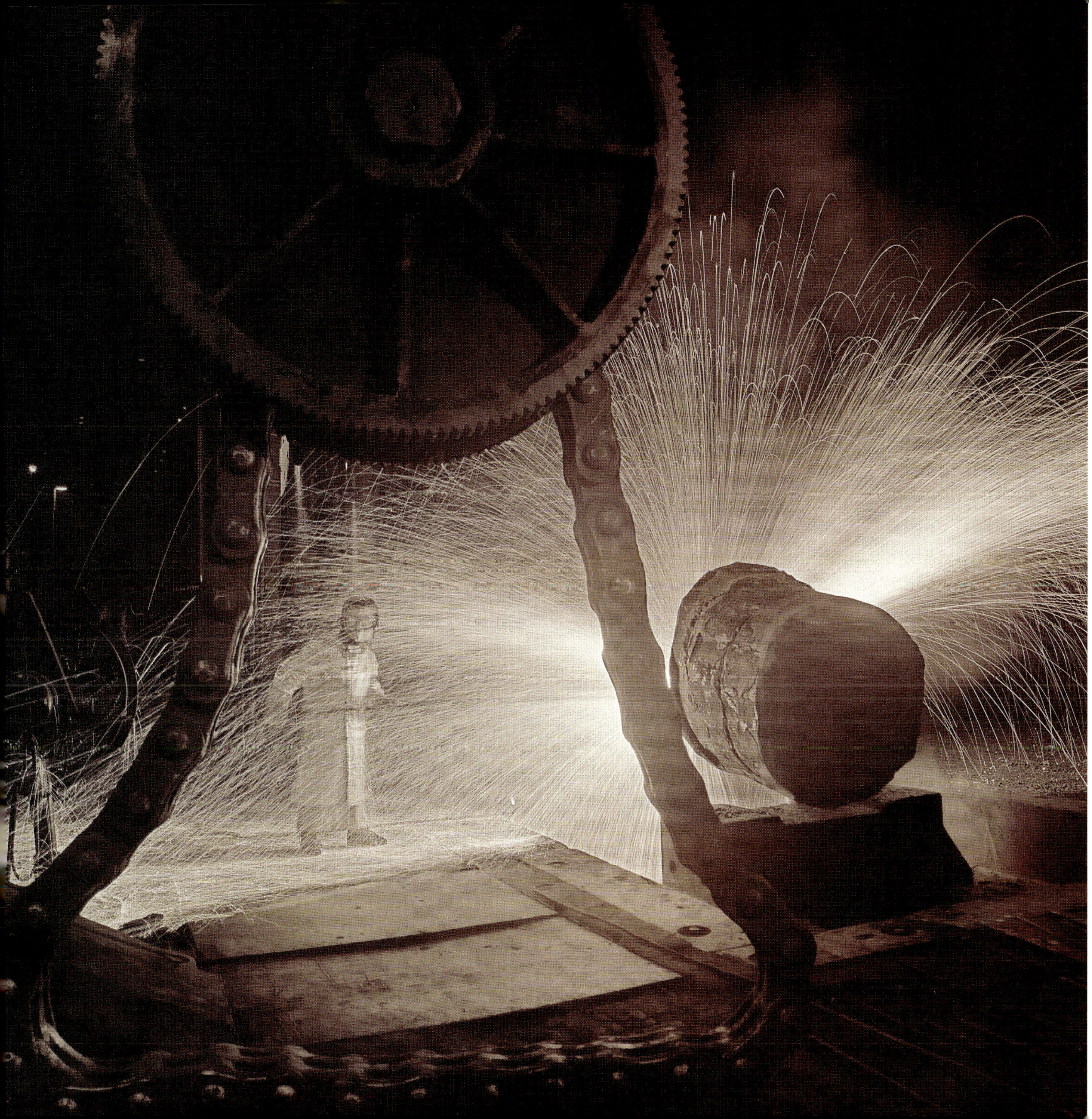

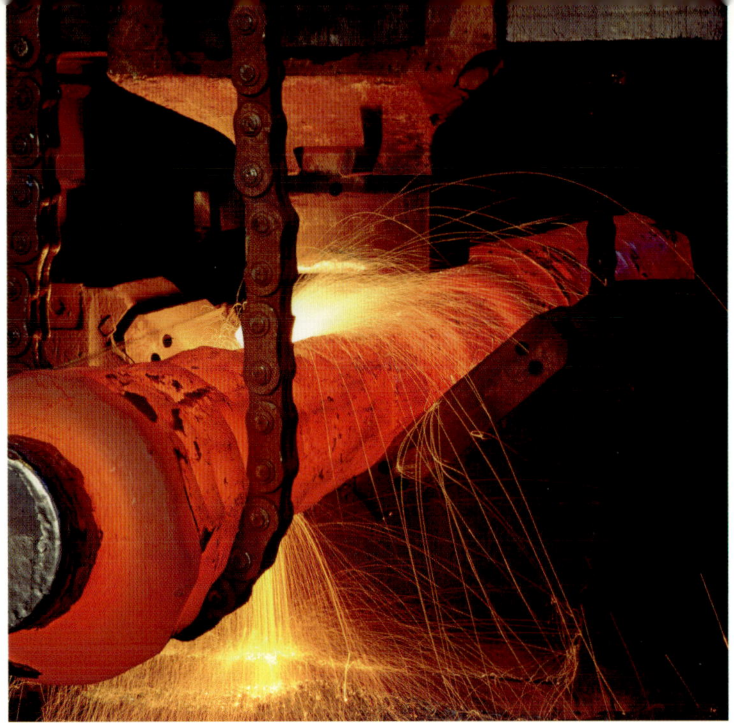

Flame scarfing

These two photos show "flame scarfing," which is the process of using pure oxygen to mend small cracks that occasionally appear during the forging process. These cracks must be repaired early to prevent propagation into the body of the forging. Forged products were used in critical applications, and the Bethlehem plant's early capabilities for forging and forming set it apart from all other competitors. As early as the 1880s, Bethlehem established a niche for heavy forgings and for customized parts. The center axle of the Ferris wheel that debuted at the 1882 World's Exhibition in Chicago was forged at the Bethlehem plant, and the Wright J5 Whirlwind engine that Charles Lindbergh used in the *Spirit of St. Louis* contained key parts that were forged by Bethlehem. The forging operation's early product lines were armored plate and ordnance, making the shops a key supplier to the U.S. military.

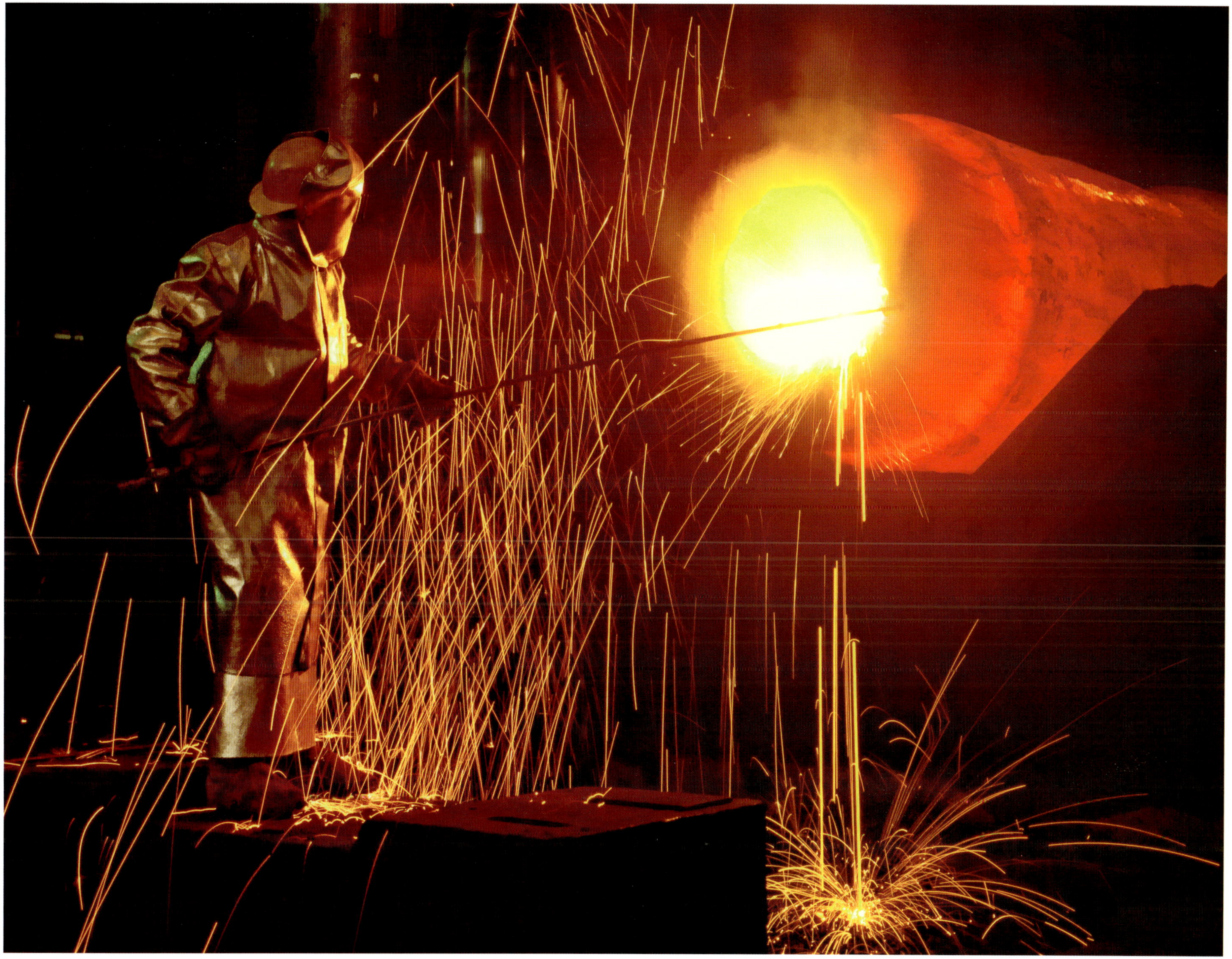

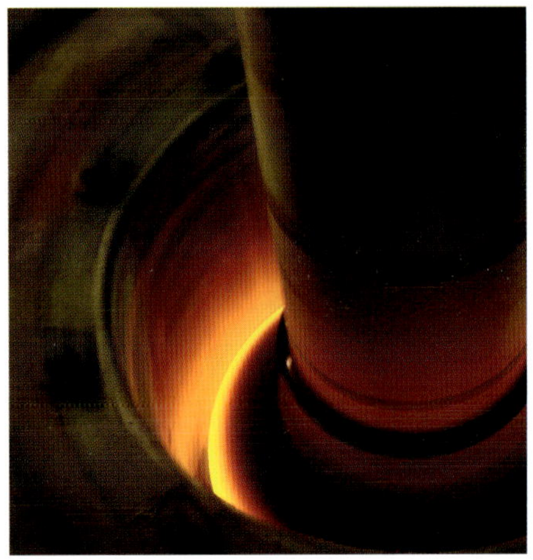

TREATMENT

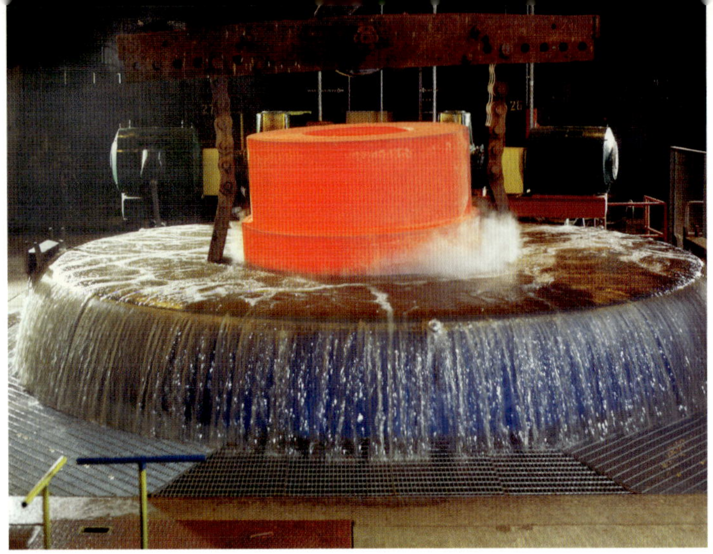

2

Water quenching

1 Rolls for the steel industry's rolling mills were one of the bread-and-butter products of the forging operations at the Bethlehem, Pa., plant. As Bethlehem Steel's empire grew, so did demand for rolls. Most other steelmakers did not have in-house capabilities to produce rolls and, therefore, contracted with Bethlehem for this mainstay product. After the rolls were forged, they required heat-treating to produce the desired structure or properties. Heat-treating involved such operations as stress-relieving, annealing, quenching, and tempering. Quenching cooled the steel at a very specific rate before it was moved to a machine shop for precision grinding and machining into its final form.

2 Water quenching was another technique used to cool the newly forged steel during the heat treating process. Bethlehem Steel drew the water used in its coking, rolling, and forging operations from nearby rivers and lakes and then cleansed the water before discharging it back into the waterways. To comply with the Clean Water Act in the early 1970s, steelmakers spent millions of dollars to install water treatment stations. Each capital expenditure increased annual operating costs, further increasing the cost structure of fully integrated mills relative to mini-mills and foreign competitors, the latter of which had less stringent, or no, government-mandated environmental regulations.

1

Air quenching steel rolls

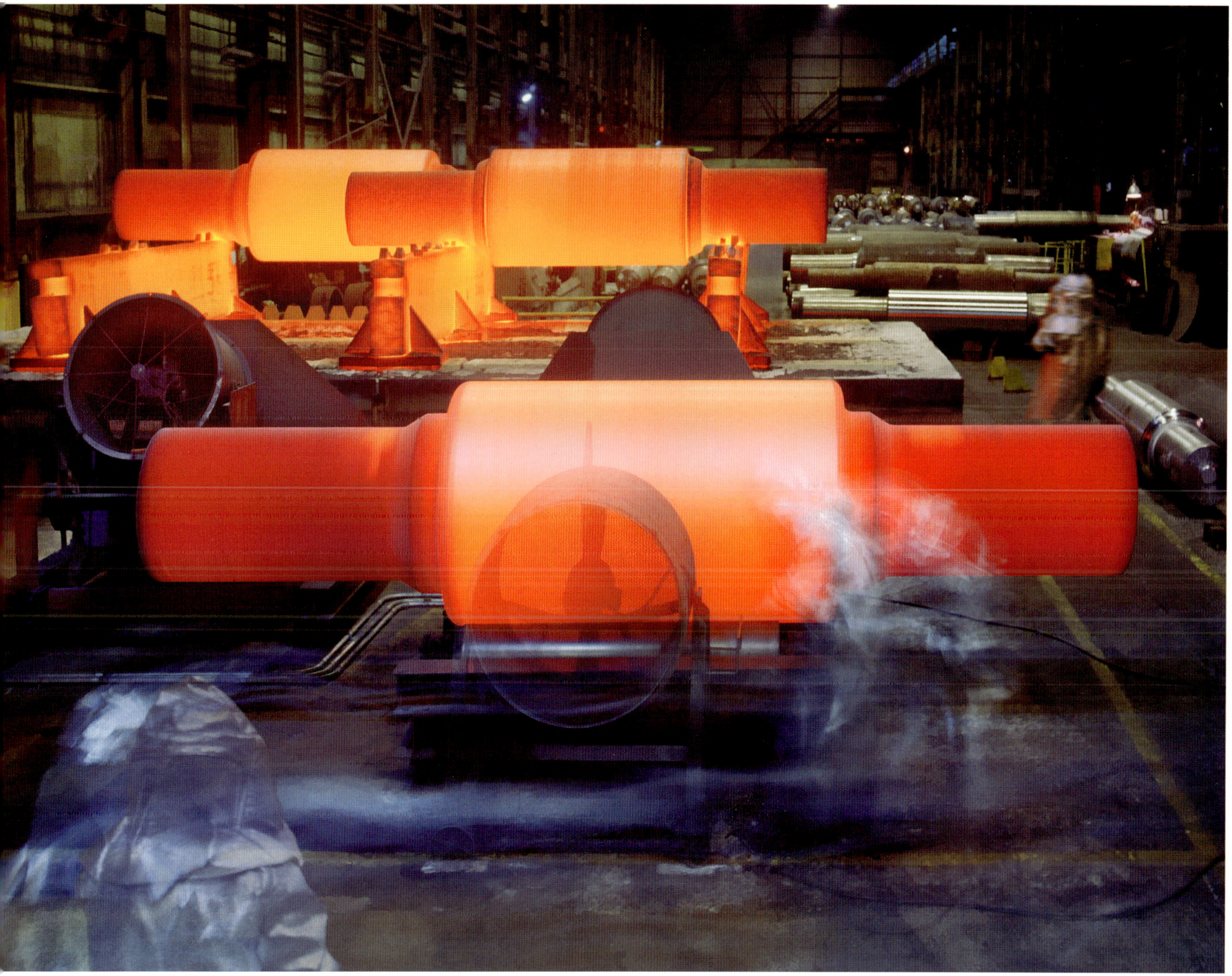

Rolls come in many shapes and sizes for both ferrous and non-ferrous mills. As important workhorses in a rolling mill, hardened steel rolls were popular products. Other roll products included forged back-up rolls and special application rolls. These rolls are receiving cryogenic treatment in a "frozen" state to create the correct properties for the customer. Cryogenic treatment chilled the rolls to -200°F. Although steel rolls were long a mainstay of the Bethlehem, Pa., plant, Bethlehem first forged various components for the military and later other industries. Demand for rolled steel was, of course, subject to the ups and downs of the cyclical domestic steel industry. When the industry was down, BethForge was able to sustain its operations with orders for other industries.

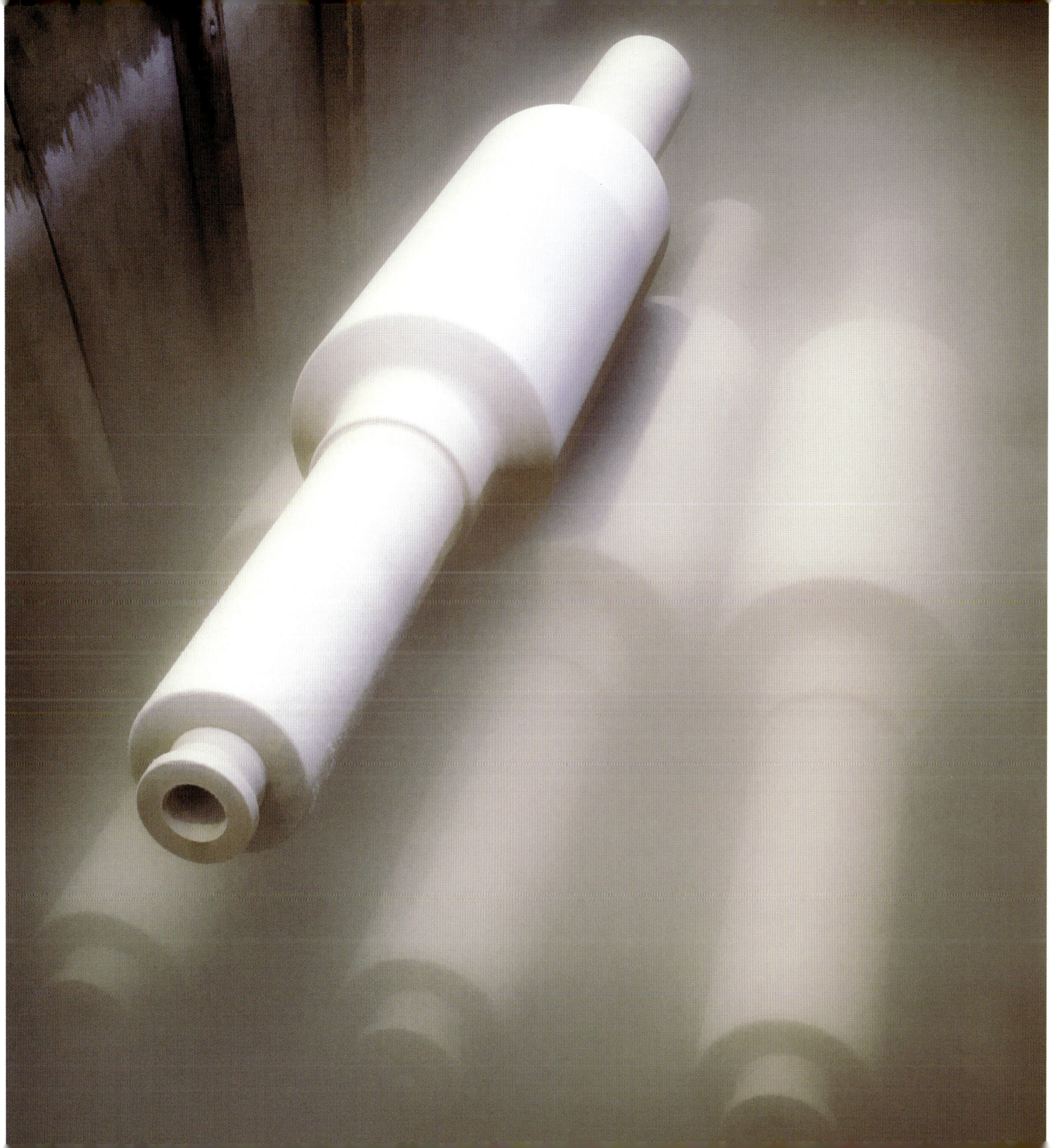

Frozen rolls

Monitoring temperature of the hardening process

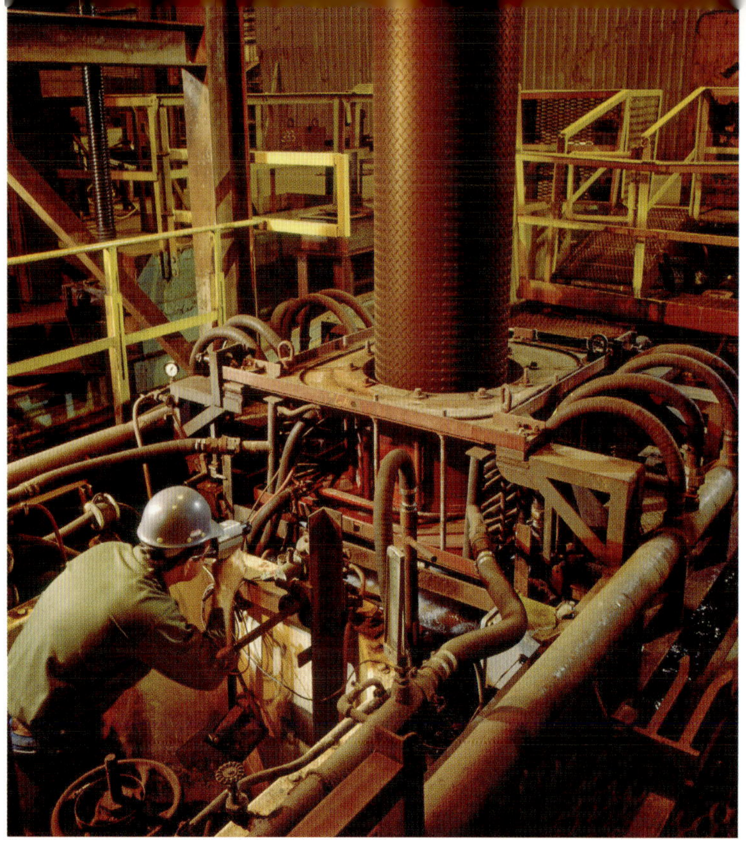

The heat-treating processes for forged products, used in the most demanding applications, included a specialized service known as "induction hardening" resulting in superior "hardness" or strength. Because of Bethlehem Steel's extensive role in both of the World Wars, the corporation had facilities to heat treat long, narrow products such as gun barrels. These facilities, known as high houses, stood up to five stories high and were as deep as they were high. Bethlehem Steel experienced its most heinous loss of life at the No. 5 high house next to the Minsi Trail Bridge, which crosses the Lehigh River. In 1941, eight plant employees died as a result of carbon monoxide poisoning when one employee, after another, a few in pairs, went to find missing co-workers and succumbed to the odorless gas. The No. 5 high house remains intact today and is expected to be maintained as part of future development of the Bethlehem, Pa., plant site.

Surface hardening by induction heating

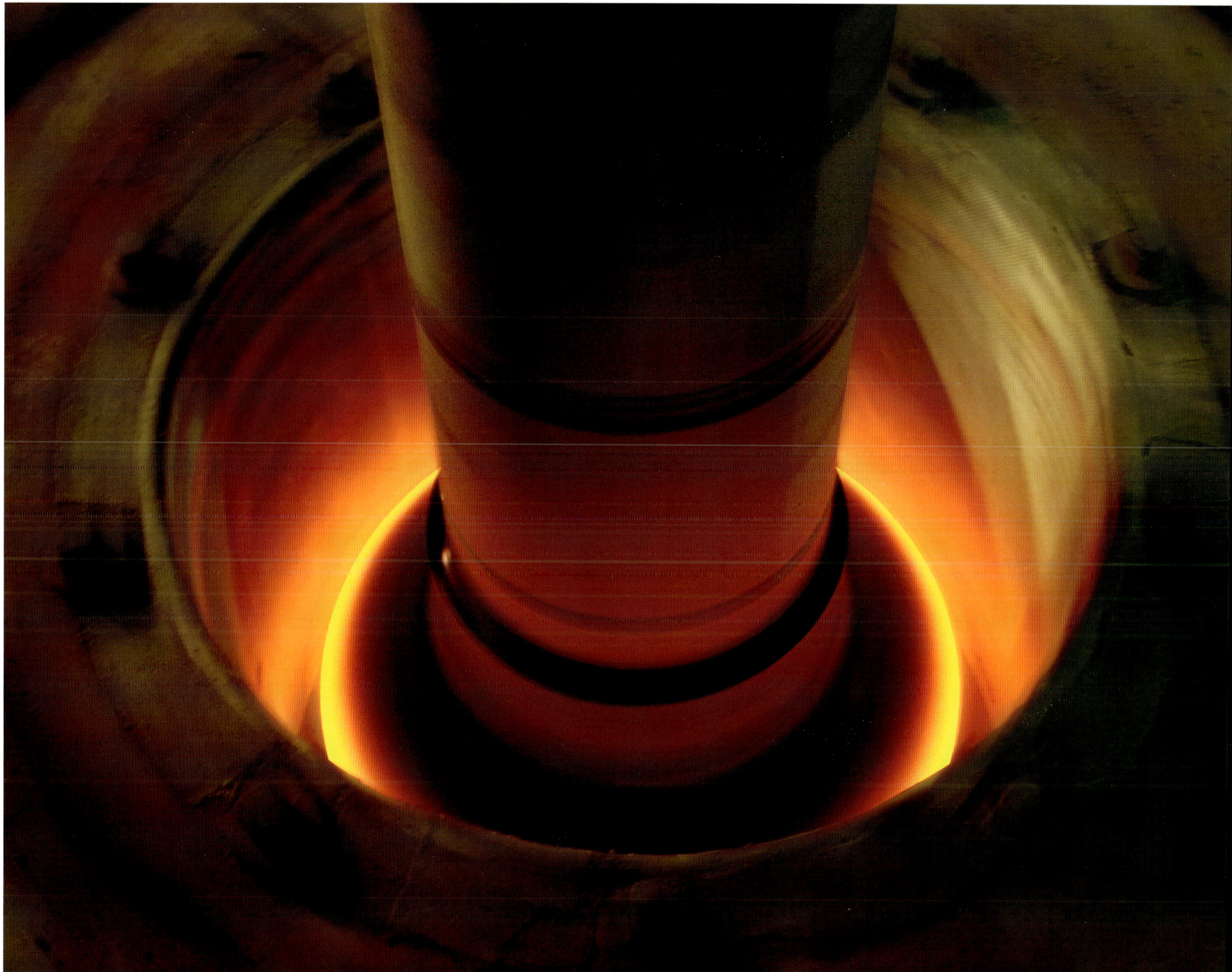

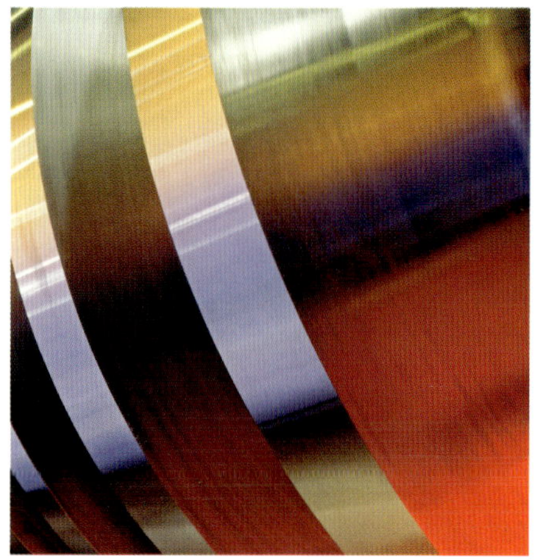

MACHINED PRODUCTS

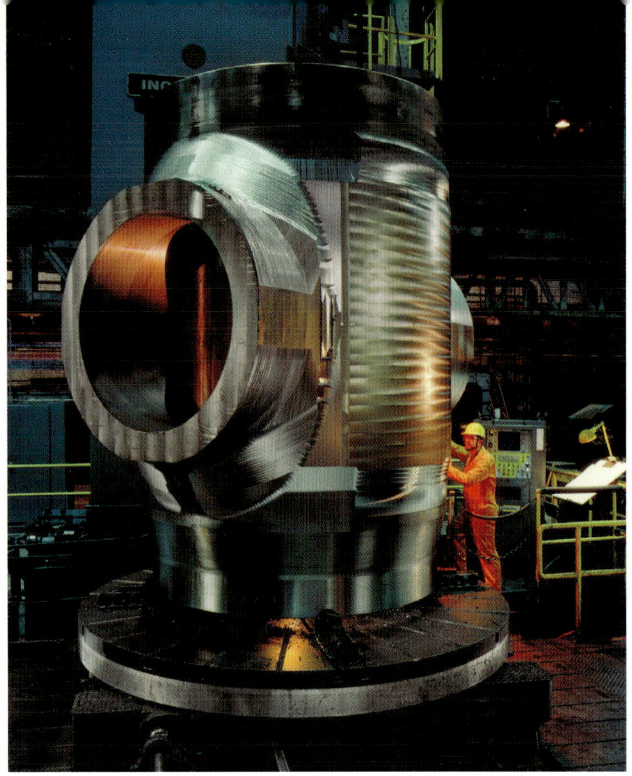

2

Shaping a nuclear reactor vessel, machine shop no. 2

1 The power generating industry was a large Bethlehem Steel customer that required exacting, and very high-value, products. The success of Bethlehem during the early years of forging in the late 1880s grew, and Bethlehem advanced its forging operations to handle the more intricate and demanding products required by the ever increasingly sophisticated needs of its customers. This gashed rotor became part of a critical rotating system in an electric power generating plant. Steam and gas producers as well as hydroelectric power plants used other high-value forgings. Lathe and trepan boring machines produced the rotor's custom shape.

2 The U.S. Navy was the longest-lasting customer of forged products made at the Bethlehem, Pa., plant. The armored plate, first made in the 1880s, convinced top Navy brass to rebuild its wooden fleet with metal, bringing the fleet in line with the naval capabilities of other industrial nations of the time. Bethlehem Steel's vast experience in providing steel for the nation's military in wartime as well as peacetime established an enduring partnership with the U.S. Navy. As the Navy began to use submarines with nuclear power in the 1950s, forged reactor shells guaranteed the containment of nuclear materials and an uncompromising performance at sea.

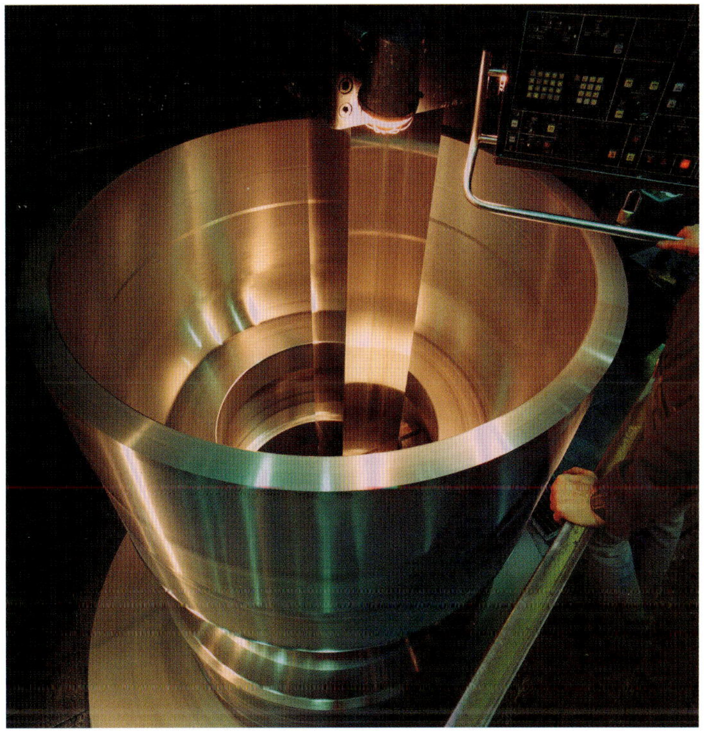

A vertical lathe

1 In addition to serving the needs of the electric, steam, and gas power generating companies, Bethlehem forged products for nuclear reactors. The extensive heat treating facilities were able to water quench sections over 18 feet in diameter with thicknesses over five feet, ideally suited for the very large products needed for nuclear applications. The heat treating provides uniform hardening for the proper balance of strength and toughness for critical pressure-vessel applications. Custom forgings were used for nuclear reactor and steam generator systems, fuel transport and storage casks, and heavy wall fabricated systems. Quality control and assurance was important in all steel products, but particularly for products used in nuclear applications. The extensive use of Statistical Process Control monitored all process parameters to maintain uncompromising quality in every step of production. A vertical lathe performed final machining, and inspections were conducted at key points during the forging's production.

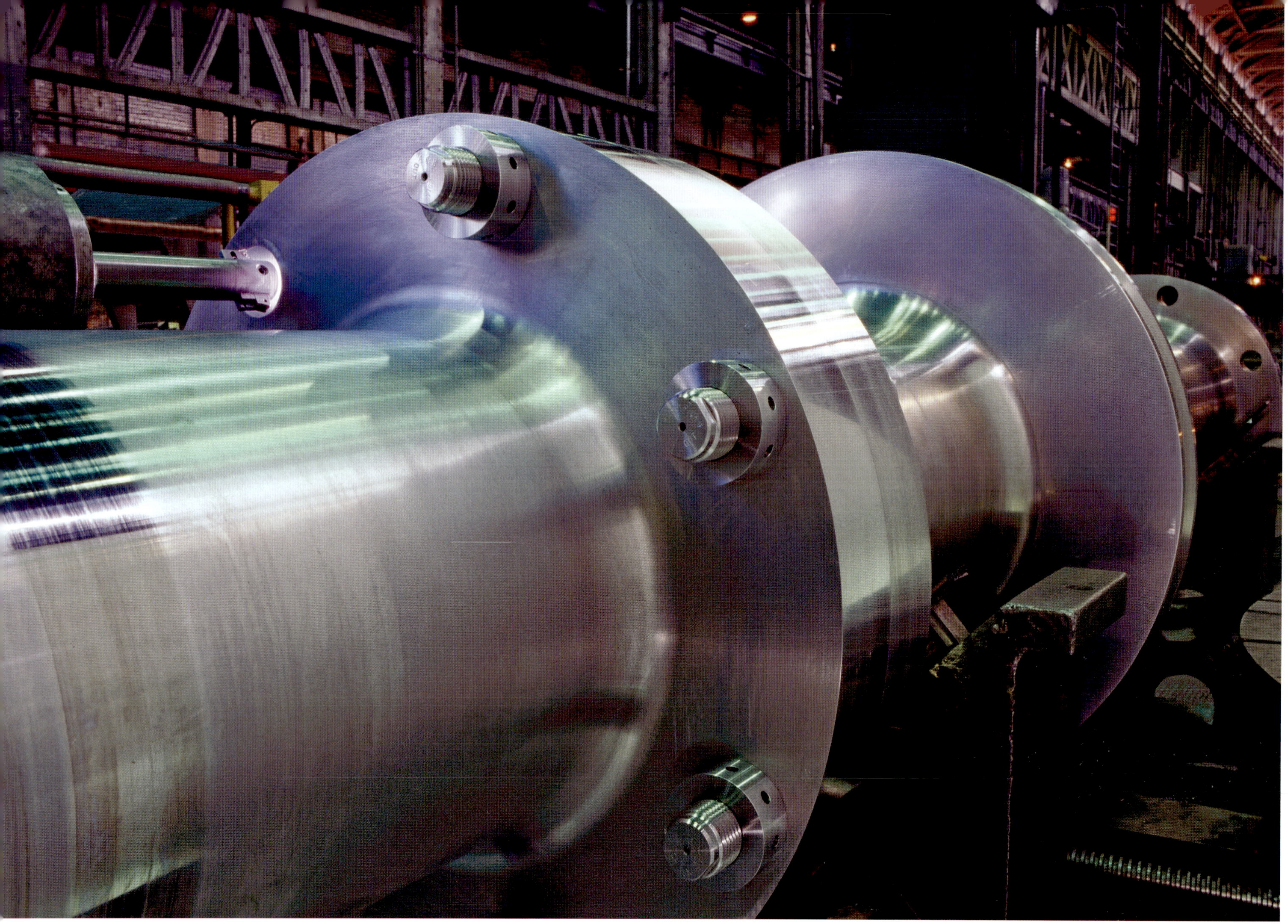

1

A propeller shaft for a large ship, machine shop no. 8

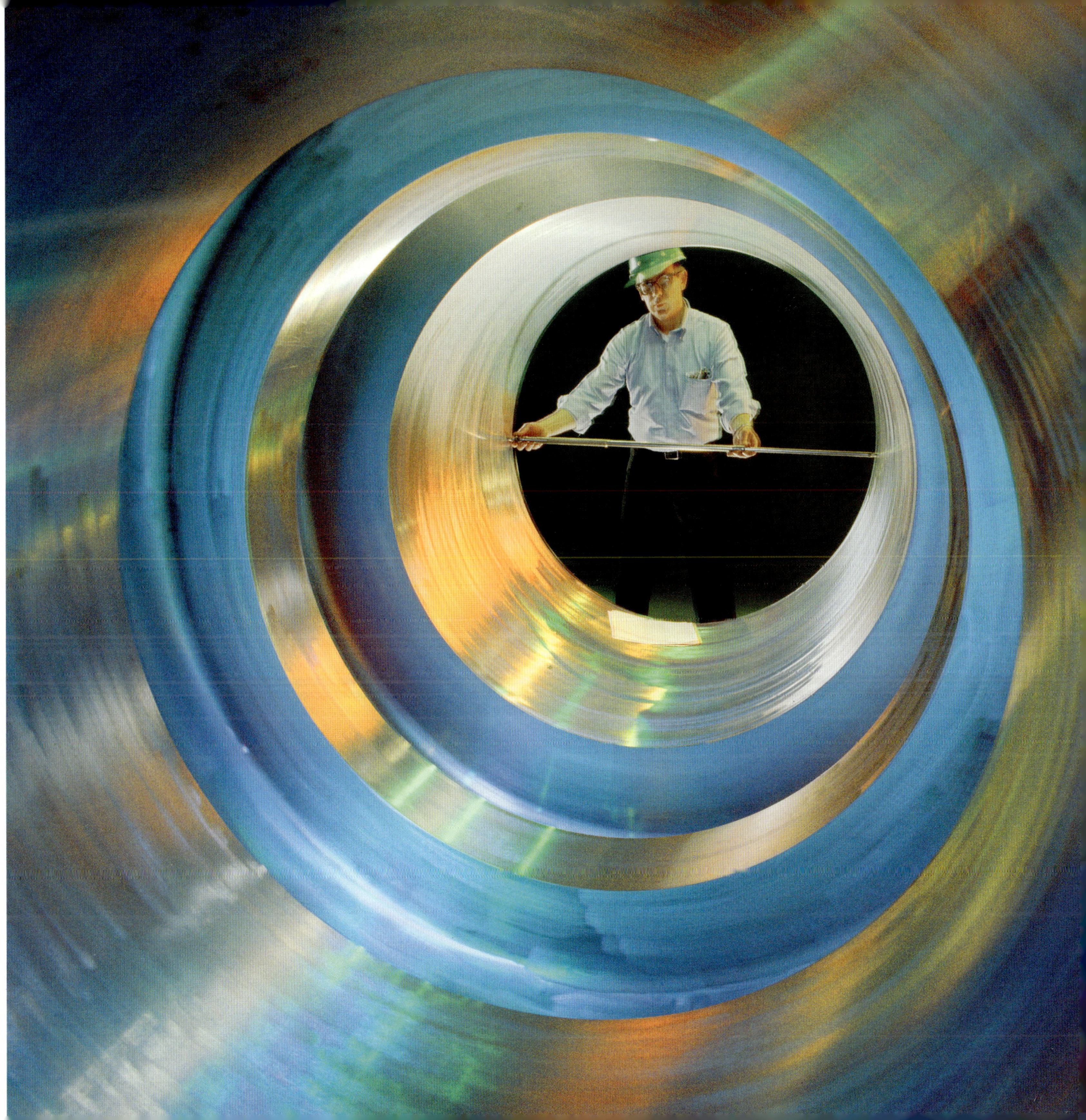

Final inspection

2

New dual head milling machine, forge specialty

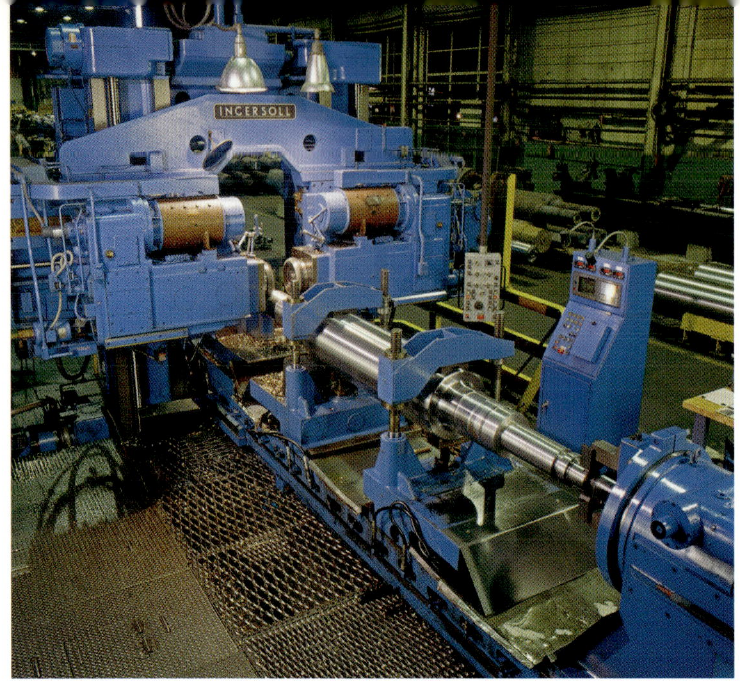

1 Because of its metallurgical, forging, and machining capabilities for forged products, Bethlehem Steel was able to develop niche markets outside of power generation and rolls, such as the market for hydraulic and pressure vessels. In 1957, Bethlehem Steel became one of the first users of vacuum degassing, an important development in the production of durable steels. Vacuum degassing removed gases such as hydrogen from electric furnace steel destined for forged products, which resulted in cleaner steels with less chance of cracking. This new process gave Bethlehem a significant leg up on its competition. The press forge, capable of handling ingots up to 400 tons, allowed Bethlehem Steel to work closely with customers needing the most precise metallurgical composition coupled with specialized forging and machining. This large hydraulic cylinder was produced for an industrial customer for use in a large machine.

2 In the early 1900s, Bethlehem Steel expanded its forged steel offerings to include both heavy and light forgings. Forge No. 3, which specialized in light forging, created smaller steel products by literally hammering, instead of squeezing, steel into a large open-die press. These smaller products were generally created for Bethlehem's many plants. In its heyday in the 1950s, the light forge division also included the central tool department, No. 6 machine shop, and the drop forge division. When the drop forge was operating, vibrations could be felt in the next building – the annex to the original headquarters building on East Third Street in South Bethlehem. Coffee cups literally "walked" on desks in offices in the northeastern corner of the annex. The forge specialty shop had lathes, grinders, and milling machines for smaller components that, in some cases, were assembled into large machinery and parts at the Bethlehem plant's weldment shop for one of the corporation's steel plants.

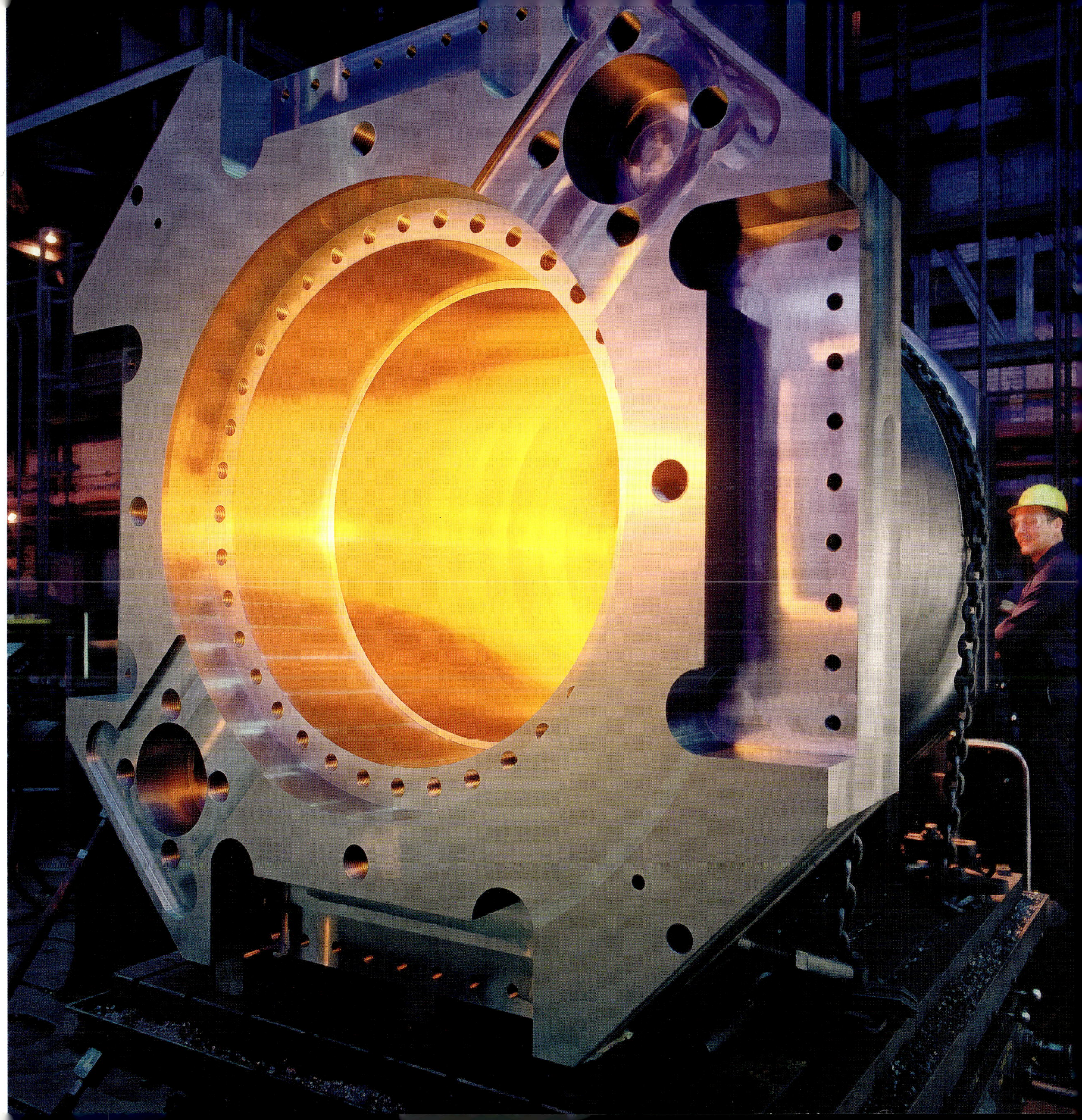

1

Hydraulic cylinder, machine shop no. 2

Beautiful scrap

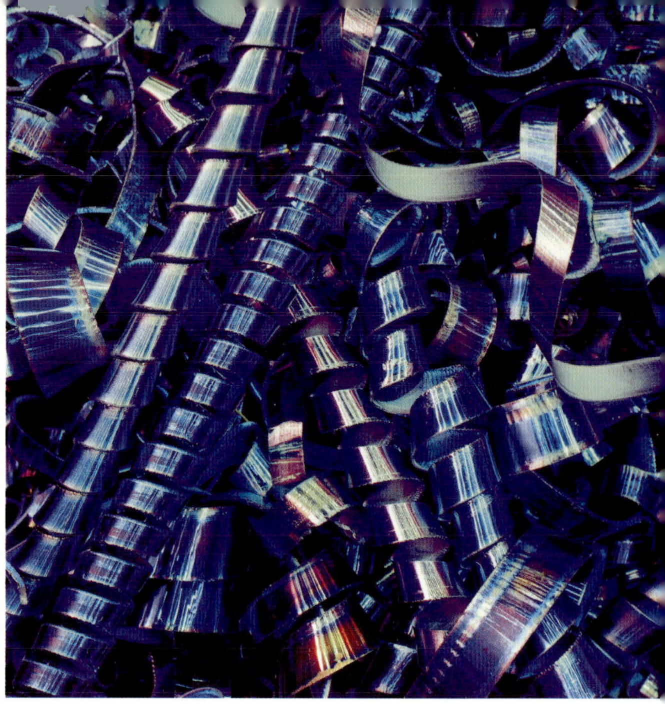

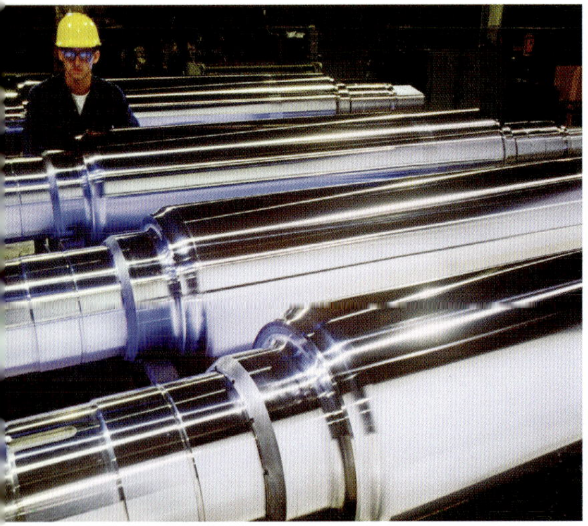

Inspecting forged rolls,
Forge Specialty Shop

Forged steel rolls for rolling mills were a mainstay of Bethlehem Steel's forging division. Its roll offerings were diverse and appealing to both ferrous and non-ferrous rolling mills. Products included hardened steel work rolls, forged steel back-up rolls, and special application rolls. In its later days when forged products became a separate business unit from the rest of the Bethlehem, Pa., plant, the sales force for forged products was composed of engineers with experience in applications, materials, and all phases of manufacturing.

In addition to rolls, the forging division also offered a wide range of products such as pinions, spindles, couplings, gear blanks, table rolls, and guide rolls for use elsewhere in rolling mills. The machining of the rolls produced razor-sharp turnings or shavings. Many South Bethlehem college students spent their summers sweeping the shavings from the floors in the No. 1 machine shop where most rolls received final machining. The turnings were returned to the electric furnace shop for remelting into molten steel.

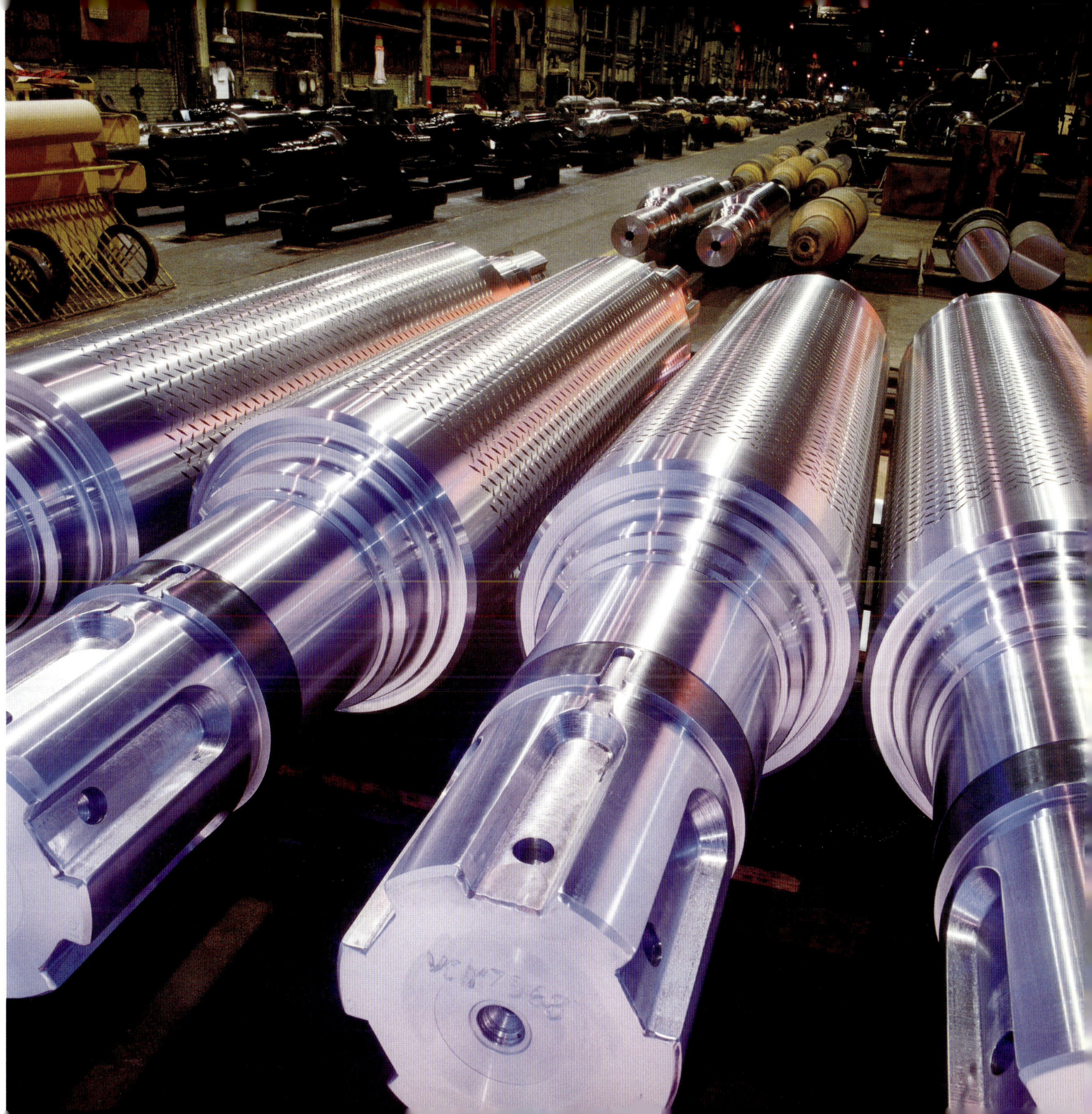

Rolls for diamond plate

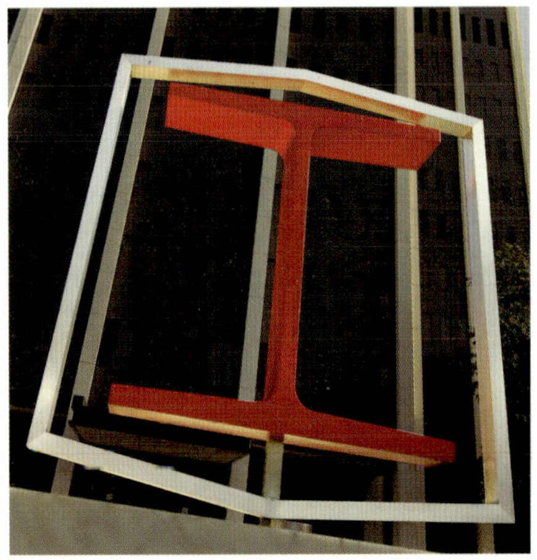

MANUFACTURED PRODUCTS

Assorted fasteners

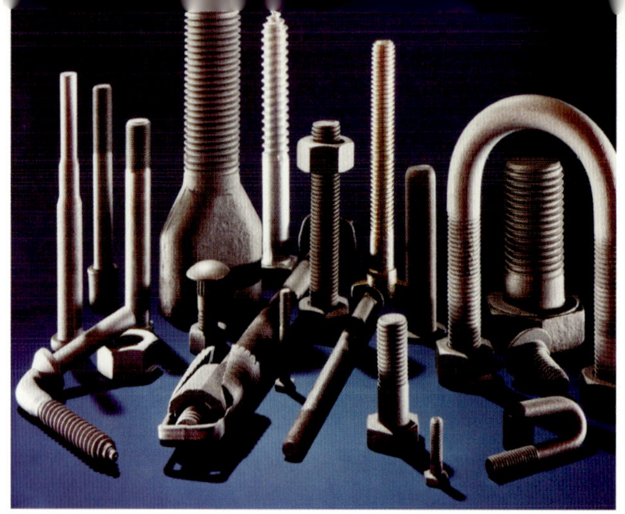

Mine roof bolts were one of the many products of the Industrial Fastener Division headquartered in Lebanon, Pa. As coal seams were mined, the bolts were placed in the roof of the new seam to prevent collapse. Bethlehem made mine roof bolts to support its coal mine operations and to sell to other mining companies, most of which were not married to a fully integrated steel producer. Other industrial fasteners included bolts for other industries as well as nuts, rivets, and spikes.

Mine roof bolts being forged,
Lebanon, Pa.

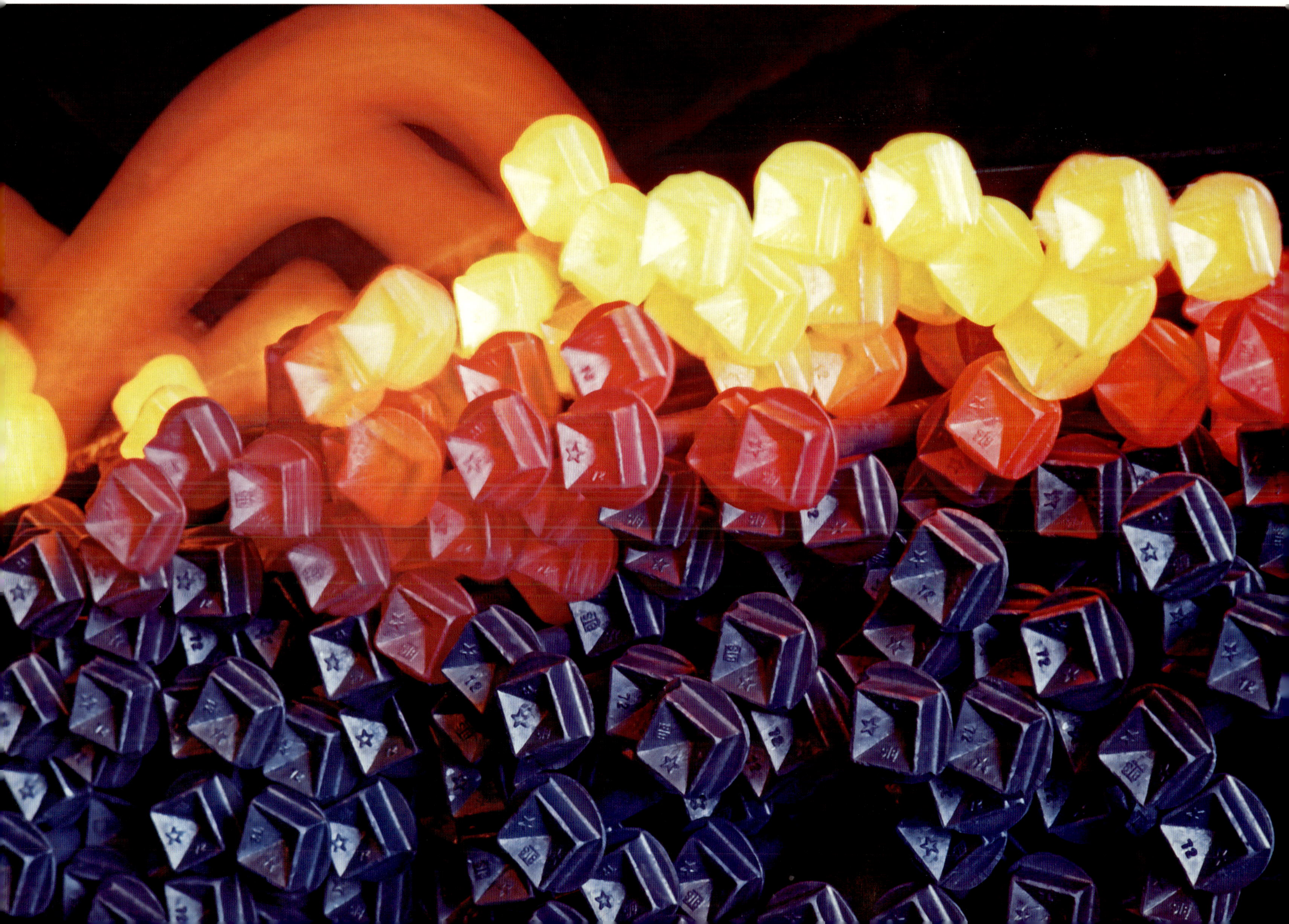

1

Five-inch diameter wire rope, Williamsport, Pa.

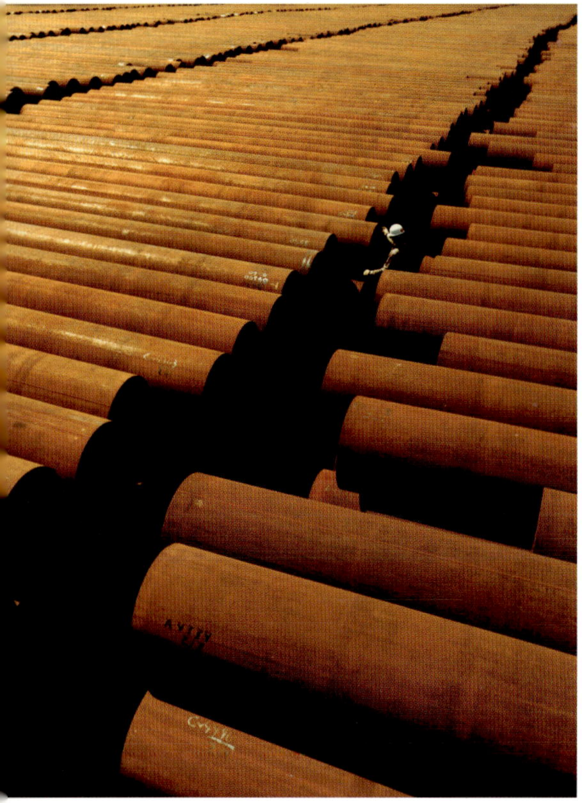

2

Miles and miles of steel pipe, Steelton, Pa.

1 Bethlehem wire rope was made in Williamsport, Pa., and incorporated into suspension cables for bridges, elevator rope, and cable systems for fabric roofed stadiums. When Bethlehem's legendary Fabricated Steel Construction Division bid on high-rise buildings and long-span bridges, it was able to offer a full range of steel products to provide nearly all of the steel components for the project. Bethlehem wire rope was used in the suspension cables for the George Washington Bridge in New York City and the Golden Gate Bridge in San Francisco as well as for elevators in most of the skyscrapers built with Bethlehem steel across the United States. Bethlehem wire rope holds the fabric in place at the Hubert H. Humphrey Metrodome in Minneapolis, one of the nation's first fabric-roofed stadiums. By the time the twin towers of the World Trade Center were constructed in the early 1970s, Bethlehem had already lost its dominance in high-rise construction. Both towers were constructed with primarily Japanese steel. However, the elevators in both towers used Bethlehem wire rope.

2 The gas shortages of the 1970s sent oil producers scrambling to develop their own pipelines to largely untapped reserves in Alaska. Bethlehem's plant in Steelton, Pa., made large-diameter pipe ideal for transmitting oil and gas over long distances. Pipe was manufactured with plate steel primarily from Bethlehem's Burns Harbor, Ind., plant and then formed in U and O presses at Steelton. The pipe shop also performed double submerged arc welding. Bethlehem inspected the welds ultrasonically to ensure permanent closure. Before shipping, the pipe was expanded to produce a perfect shape and improve the mechanical properties of the steel. Bethlehem was a major supplier to Alaskan pipelines built in the 1970s and 1980s to lessen America's dependence on foreign oil.

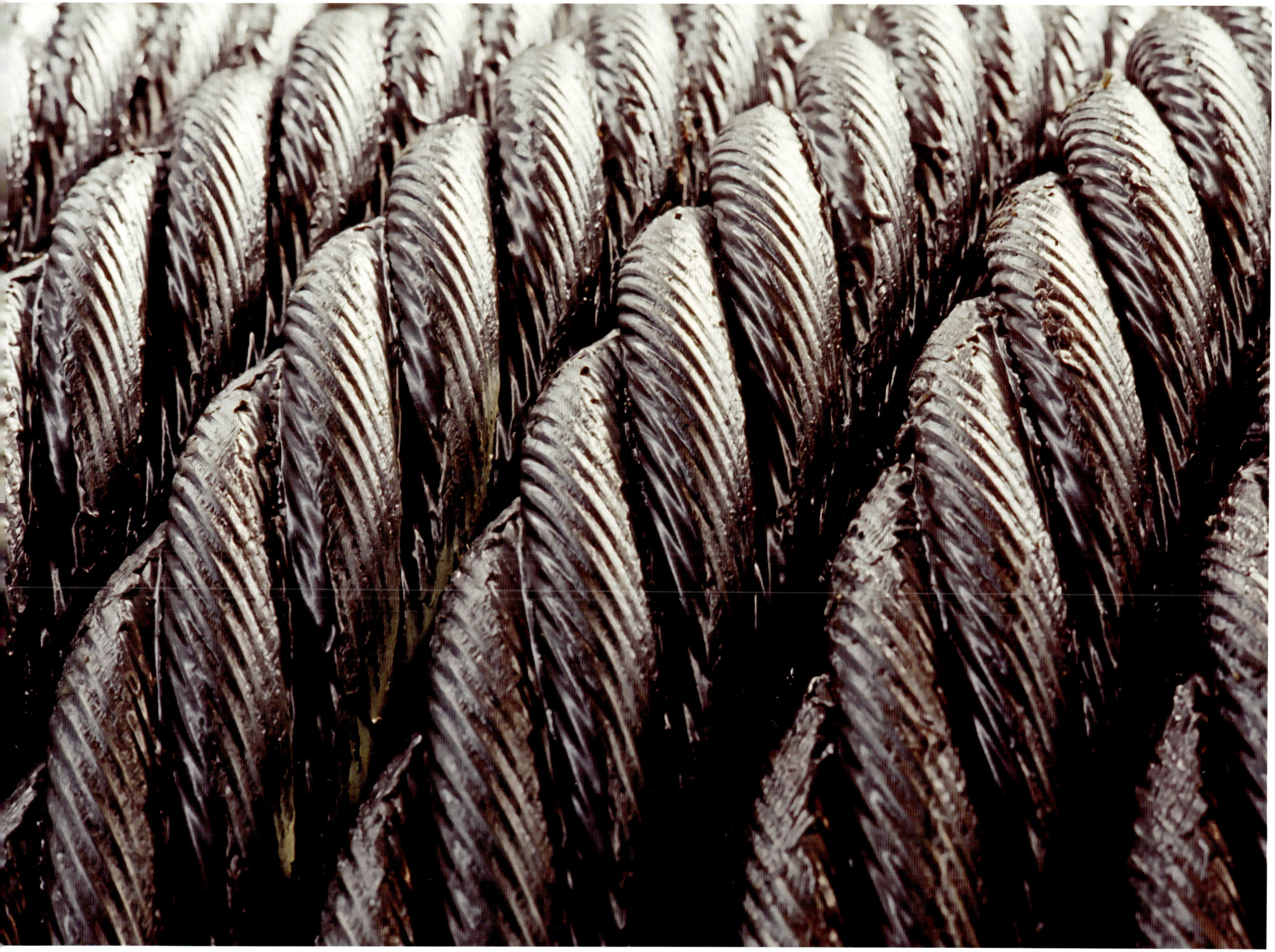

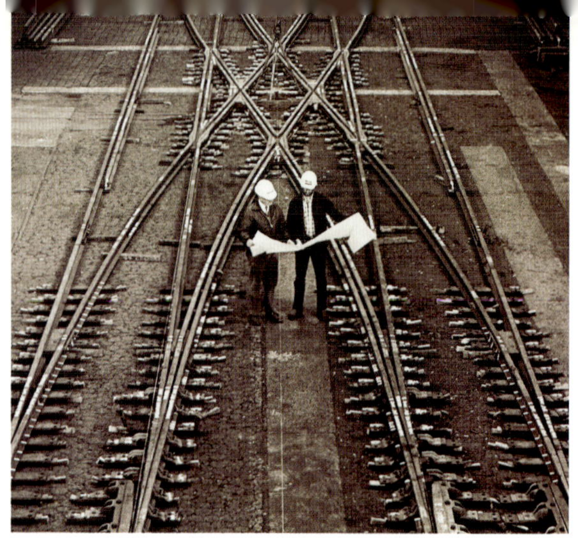

1

A double crossover for the New York City Subway

1 The only Bethlehem long product that continues to be made today is rail, which is noteworthy because rail was the first product of the company's predecessor, the Bethlehem Iron Company, which first rolled iron rails in 1863. By the early 1870s, steel proved to be more durable than iron, and Bessemer furnaces were installed in Bethlehem to make steel. In 1915, Bethlehem acquired the Maryland Iron and Steel Company and its subsidiary, the Pennsylvania Iron and Steel Company, which has operated since 1863, in Steelton, Pa, about 90 miles west of Bethlehem outside of Harrisburg, Pa. In the 1970s, Bethlehem closed its Lackawanna, N.Y., rail mill and consolidated rail production at the Steelton plant. Steelton had the operating and cost advantages of electric (scrap-based) steelmaking and a continuous bloom caster, which had been installed in 1983. The Steelton plant also had a shop for the production of frogs and switches, the precise fabricated metal plates that diverted rails seen here above the men in the photo. Although the frog and switch shop was closed in the early 1990s, its products were sold at premium prices because of their intricate nature. Today, Steelton, now owned by Mittal Steel, continues to roll a variety of rails for long-haul, short-line, and rapid-transit systems. At the end of Bethlehem Steel in May 2003, Steelton was providing rails to rebuild the portion of the New York City subway system destroyed during the September 11th attacks in 2001.

2 Since 1901, coal-carrying rail cars have been made in Johnstown, Pa., first by Cambria Steel Company and, beginning in 1923, by Bethlehem Steel. The corporation's Freight Car Division in Johnstown, Pa., used a variety of products, including plates and structural shapes from throughout the corporation and bars from Bethlehem's Johnstown-based Bar, Rod, and Wire Division. In 1978, Bethlehem introduced the BethGon (short for Bethlehem Gondola) Coalporter coal car, which was very popular with customers who needed coal delivered to their facilities. Responding to customer requests for a lighter but durable car, in 1986 Bethlehem introduced the aluminum BethGon Coalporter. The Freight Car Division was sold in 1991 and continues to operate today as FreightCar America.

2

Coal Porter cars from Johnstown, Pa.

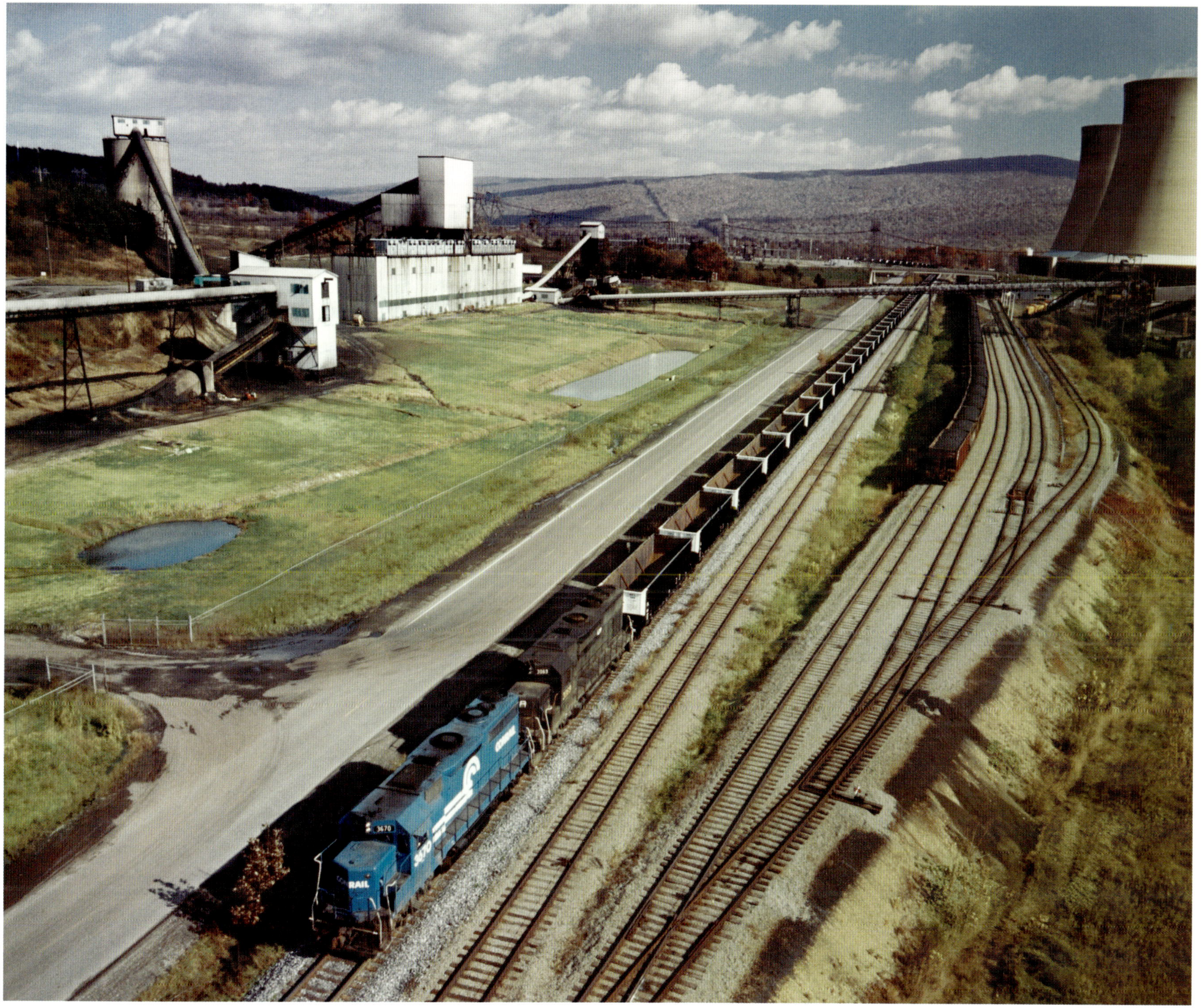

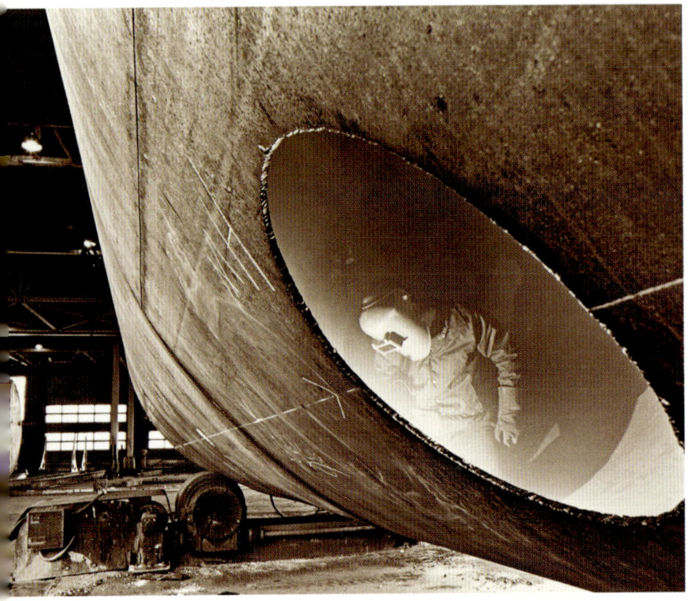

1

Welding a tank, Baltimore, Md.

1 With the advent of the suburb after World War II, the demand for underground storage tanks took off. Steel tanks held heating oil for suburban homes and gas for filling stations that popped up all over America as communities expanded and people took to the roads. Bethlehem's Buffalo Tank Division met that new need and expanded on the corporation's expertise with other types of underground and containment vessels used in military and industrial applications. Once the early 1980s rolled around, demand for Buffalo Tank products was fading, and the division was put up for sale. Bethlehem placed its divisions engaged in manufactured products in the steel-related group, dubbed by cynical employees as the "expendable group." By the end of the 1980s, Bethlehem was no longer engaged in the myriad manufactured products that supported its long-time role as a complete "supermarket of steel."

2 The man on the ladder is sandblasting the rudder of a ship in the Hoboken shipyard directly across the Hudson River from midtown Manhattan. Bethlehem's preeminence in shipbuilding was instrumental to the naval campaigns of the Allied Forces during both World Wars in the early decades of the 20th century. In 1943, long-time Bethlehem President Eugene Grace pledged a ship a day to the Roosevelt administration to help the Allies win the war. In all, Bethlehem's vast empire of shipyards produced nearly 1,200 ships. During World War II, the corporation was the largest shipbuilder in the world. Bethlehem's shipbuilding yards were located in Boston, Mass.; New York, N.Y.; Baltimore, Md.; San Francisco, Calif.; San Pedro, Calif.; and Beaumont, Texas. The Hoboken, N.J., ship-repair yard repaired and modified ships for combat. In the 1980s, realizing that America's shipbuilding days were over, Bethlehem sold most of its shipyards. Acquiring firms dismantled most of the shipyards and sold the water-front properties to condominium developers. Bethlehem's Marine Construction Division at the Sparrows Point, Md., shipyard fabricated the tunnel sections used in the "Big Dig" for the Boston Harbor Tunnel. That yard was sold in the mid 1990s.

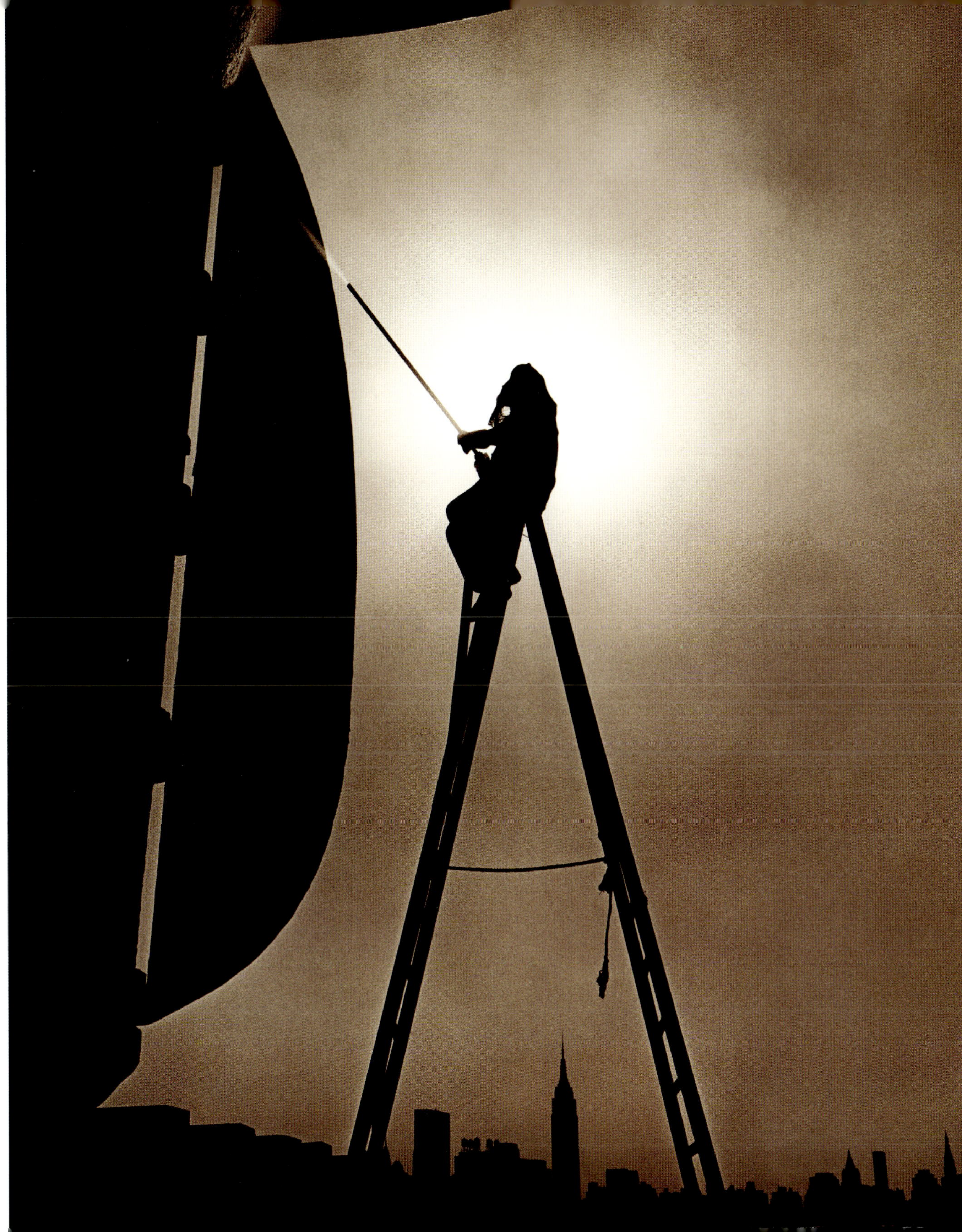

2

**Sandblasting a ship,
Hoboken, NJ.**

1

Cutaway view of a new ship, Philadelphia, Pa.

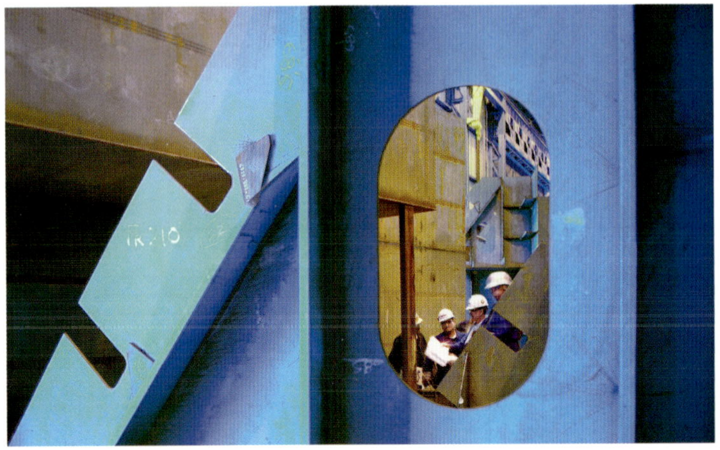

2

Kvaerner inspection team

1 The shipbuilding industry in the United States had been declining for years when, in the late 1990s, Kvaerner Shipyards of the Netherlands bought the shuttered Philadelphia Navy Yard. That purchase closely followed Bethlehem's acquisition of Lukens Inc. in nearby Coatesville and Conshohocken, Pa., making Bethlehem the largest producer of plate products in the United States. In 1999, Bethlehem successfully became the sole plate steel provider to the shipyard. It was at the Kvaerner yard that the hull of the *U.S.S Cole*, severely damaged during the 2000 terrorist attack in the Persian Gulf, was repaired with Bethlehem plate products from Burns Harbor, Ind., and Coatesville, Pa.

2 To ensure a perfect fit and performance while at sea, naval vessels were extensively inspected at all phases of production. Exacting fabrication and outfitting work was completed in dry docks under the watchful eyes of marine engineering specialists. During World War II, when all of Bethlehem's steel was destined for the military and much of it for the U.S. Navy, hundreds of Navy employees were stationed at the supplying steel plants and at all shipyards to monitor production, fabrication, and construction. From the metallurgical testing of the steel to final outfitting, the Navy made certain that Bethlehem adhered to the Navy's exacting standards. During the war years, Bethlehem Steel was the largest contractor to the military followed by General Motors.

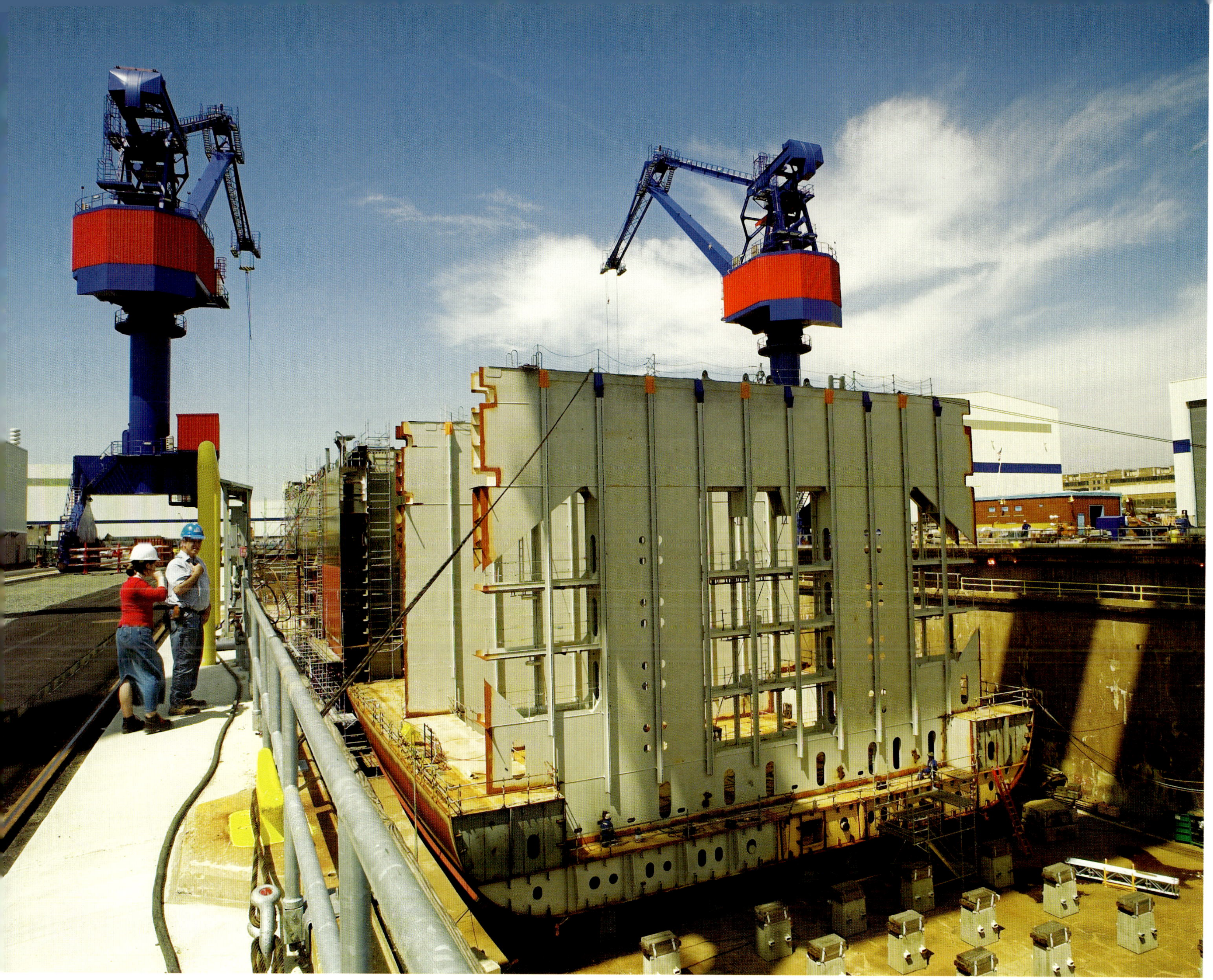

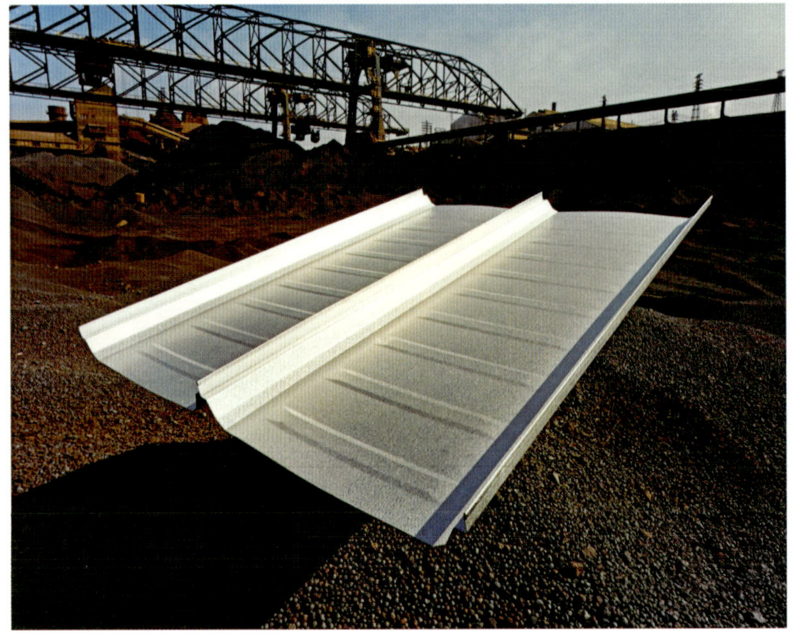

Brochure image made in Bethlehem, Pa. ore yard

In the 1960s, research engineers, along with mill personnel at Sparrows Point, Md., developed Galvalume sheet steel, which provided superior corrosion resistance. First produced in 1972 at Sparrows Point, cold-rolled steel was coated with zinc and aluminum in a continuous hot-dip process. In the late 1970s, Bethlehem embarked on a huge advertising campaign that targeted metal building and roofing manufacturers and showed the sheets in hostile environments. Pictured to the right is a sheet "floating" above the ore yard at the Bethlehem, Pa., plant where thousands of tons of iron ore pellets were stored and moved by the huge overhead cranes and their clam-shell buckets to waiting ore buggies. Galvalume was a huge asset to Bethlehem Steel as the corporation began to emerge from the depths of the industry-wide recession and restructurings of the 1980s. The product was highly regarded by metal building manufacturers and for higher-end uses in residential roofing for its corrosion resistance and flawless finish. Bethlehem maintained the patent and world-wide licensing of the coating for many years.

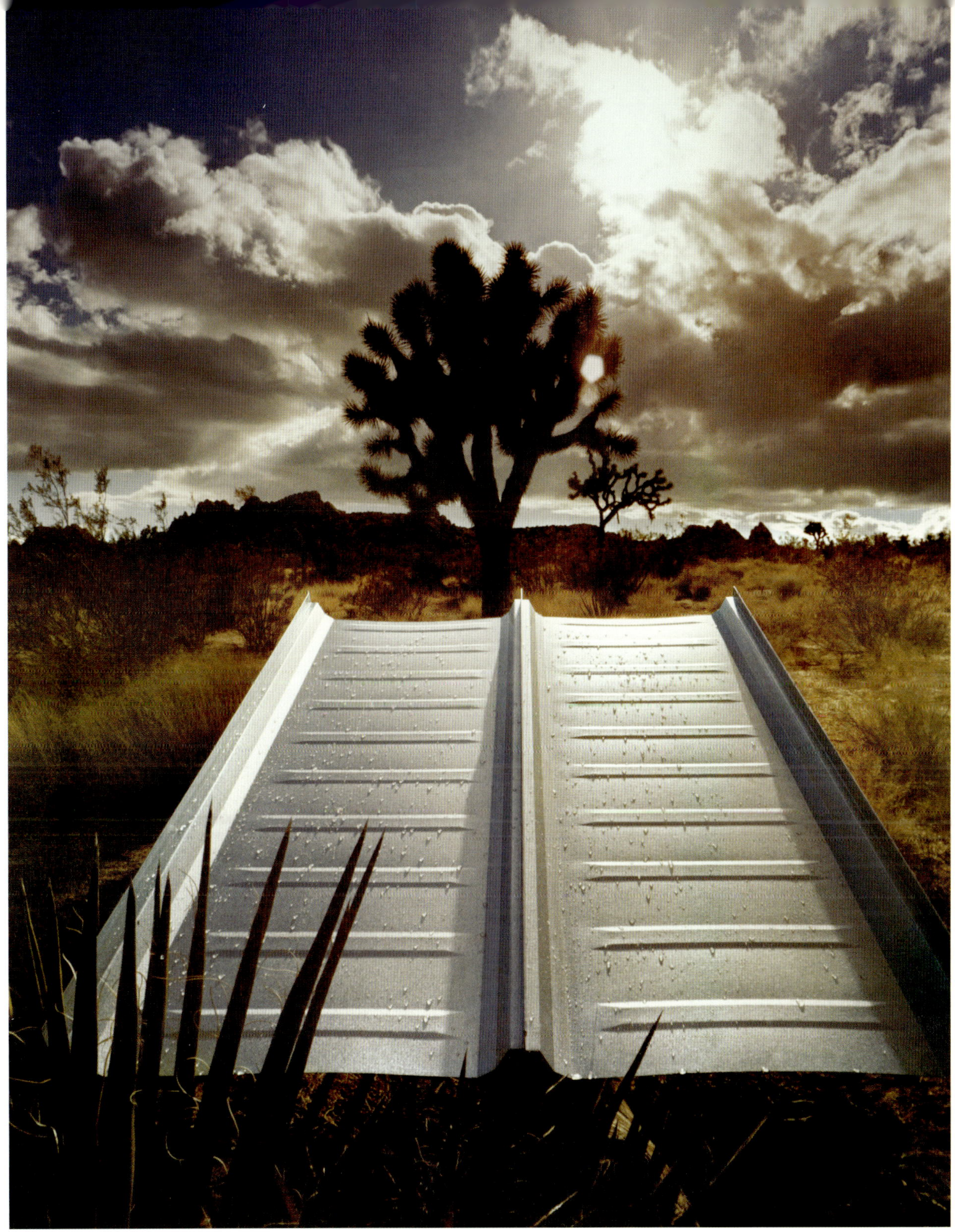

Advertising poster image
for Galvalume roof panels

Looking west from Homer Research Lab observation tower

Bethlehem opened Homer Research Laboratories in 1959. Homer Labs provided Bethlehem with an unparalleled facility to perform extensive research and product development. The mountain-top campus overlooking the City of Bethlehem contained 10 buildings, one of which included miniature versions of a steelmaking furnace and rolling mill. At its height, Homer Labs employed more than 1,000 researchers, many of whom had doctoral degrees in sciences, such as metallurgy, essential to the company's varied products and processes. In the later years of the company, Homer Labs placed particular emphasis on the development, and further refinement of, sheet products, particularly coated ones, which constituted the company's highest-value products. This large photo is taken from the observation tower of the main building, which housed a library, a small auditorium, dining room and, of course, research executive offices. The smaller photo shows a test for formability to meet the ever-increasing requirements of the automotive industry. During the "fire sale" of assets to stave off bankruptcy in the mid 1980s, Bethlehem sold most of the buildings to Lehigh University, its neighbor to the north side of the mountain on which the research center was located.

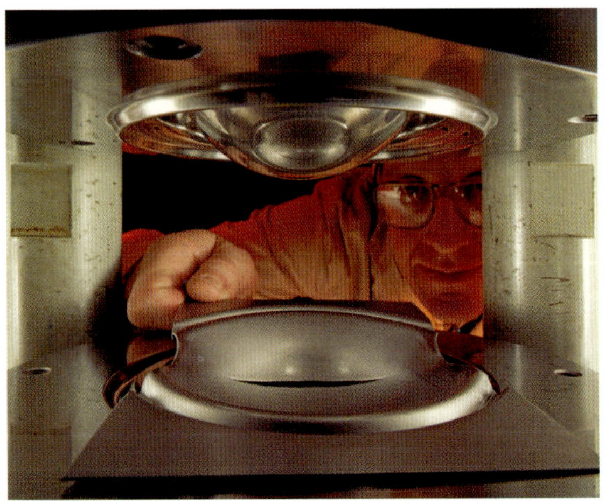

Formability testing

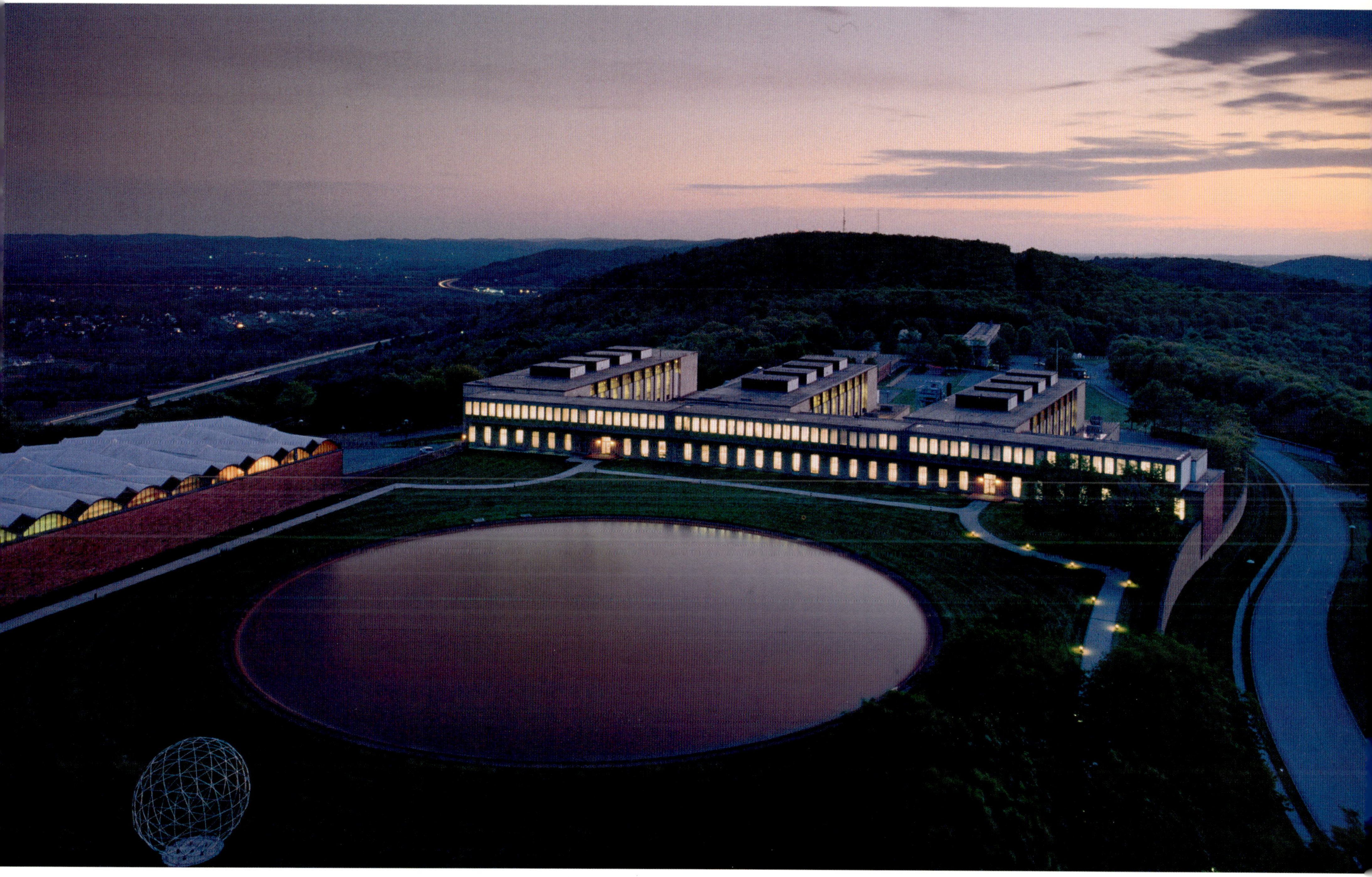

Martin Tower

Opened in 1972, Martin Tower, named after retired chairman Edmund Martin, was the then-new headquarters of Bethlehem Steel Corporation. Built as a cruciform, reminiscent of the iron foundry (probably purely by coincidence) at the Bethlehem plant where steel rails were rolled, the shape of the building provided the maximum number of corner offices. Only vice presidents and above had the southeastern corner offices, which came with private rest rooms. Opulent but functional when new, Martin Tower boasted fine furnishings, original artwork, plush carpeting, and excellent food services. Handmade tapestries from England depicting the basic oxygen furnace process hung in the ground-floor library and the executive level. The board room had a reception area for executive assistants who staffed the phones and presenters who waited to be called in at a moment's notice. If you held a position high enough to be granted an underground garage space, you were required to park American-made cars. Later, this requirement was modified to include foreign cars made in America or made with Bethlehem content. At the end of the company's life and with an "outsider" at the helm, Bethlehem lifted the no-foreign-car ban, and in came the high-end imported cars. However, the building fostered the dreams and hard work of thousands of people during its lifetime. Today, Martin Tower is expected to be converted into condominiums; penthouse units, with their spectacular views, are expected to command seven-figure prices.

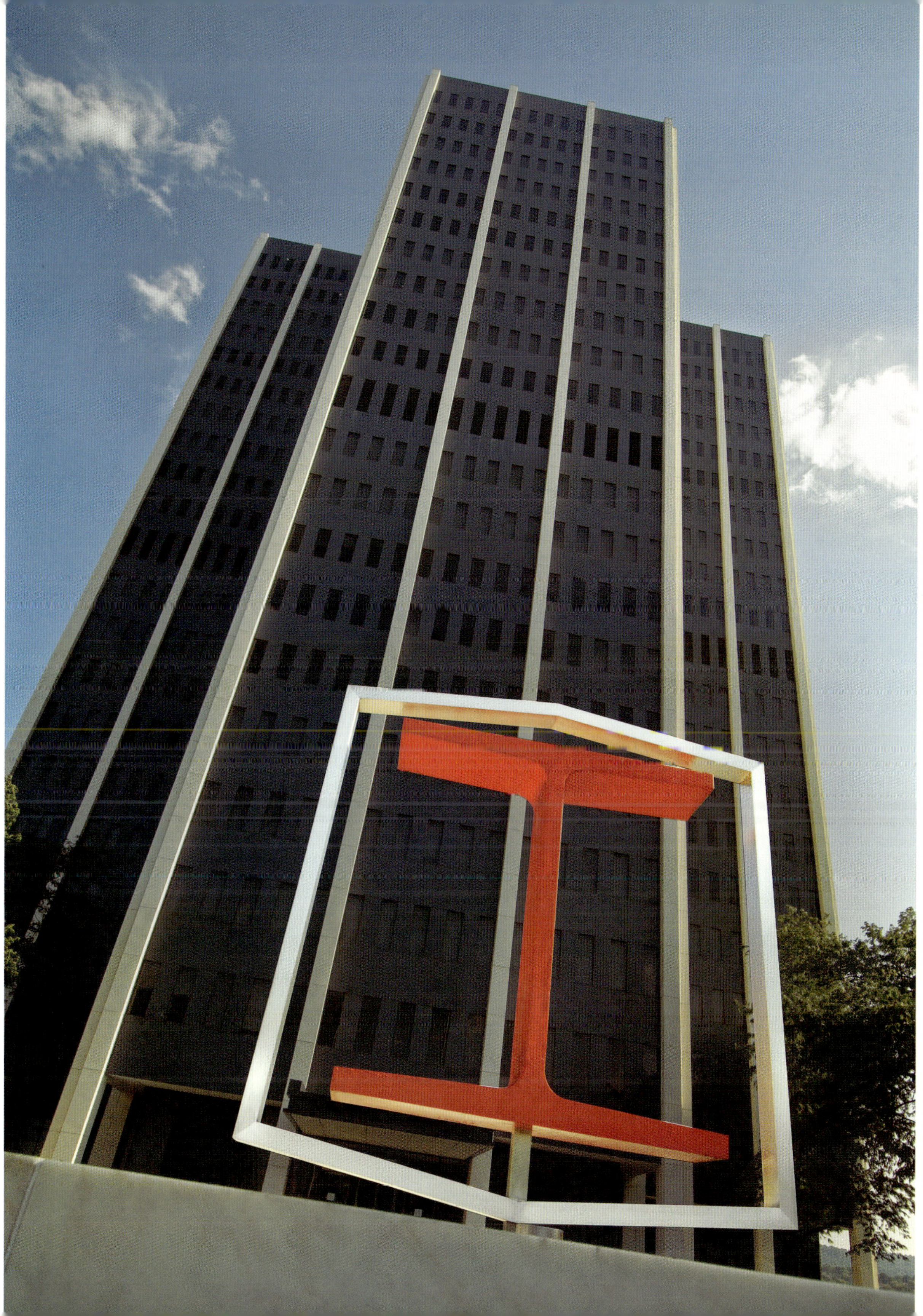

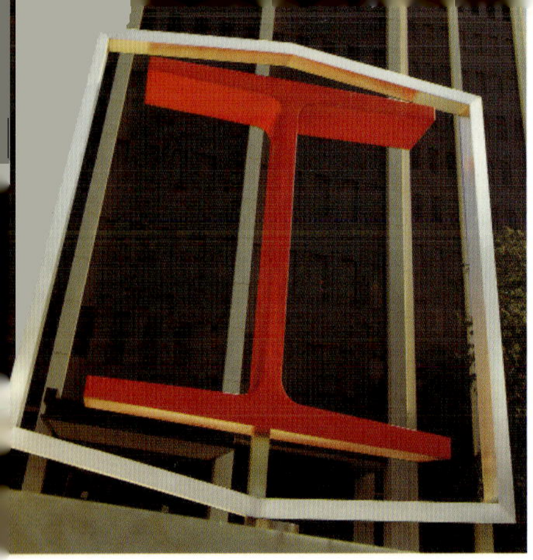

TIMELINE

Timeline of the Life of an American Icon: Bethlehem Steel Corporation

1857 Saucona Iron Company, Bethlehem Steel Corporation's predecessor, is established along the south shore of the Lehigh River in south Bethlehem.

1859 Bethlehem Rolling Mill and Iron Company is established as the successor to Saucona. A patent is received for a rail rolling mill.

1861 The company is reorganized as the Bethlehem Iron Company.

1873 Steelmaking begins in Bethlehem with a Bessemer furnace to produce more durable steel rails.

1886 Forging hammers, presses and a large machine shop are added to the Bethlehem plant to produce ordnance and armor plate for the U.S. Navy.

1899 Bethlehem Steel Company is formed with the Bethlehem Iron Company as its only asset.

1904 Charles M. Schwab, formerly of U.S. Steel, forms Bethlehem Steel Corporation.

1906 Bethlehem invests $5 million of technology by Englishman Henry Grey to build a rolling mill for wide-flange beams.

1908 Grey mill opens to usher in the age of skyscraper and long-span bridge.

1916 Bethlehem doubles its steelmaking capacity with purchases of Pennsylvania Steel Company in Steelton and the Maryland Steel Company at Sparrows Point. Bethlehem also purchases the American Iron and Steel Manufacturing Company in Lebanon and various mining properties in Pennsylvania and West Virginia.

1917 Bethlehem Steel Corporation becomes the third-largest industrial company in the United States after U.S. Steel and Standard Oil of New Jersey.

1922 Bethlehem buys Lackawanna Steel Company near Buffalo, N.Y. and iron ore mines in the Great Lakes region.

1923 Pension, employee savings and stock ownership plans are established as Bethlehem buys most of the Midvale Steel Company of Philadelphia and the holdings of the Cambria Steel Company, including its steel operations and coal mines in and near Johnstown, Pa.

1930 The United States' west coast is being developed, and Bethlehem buys the Pacific Coast Steel Company in Seattle and Southern California Iron and Steel Company in Los Angeles.

1941 The Steel Workers Organizing Committee is recognized by Bethlehem Steel as the employees' representative, leading to the rise of the United Steelworkers of America.

1959 The United Steelworkers of America go on strike for 116 days. The industry agrees to wage and pension demands. Bethlehem Steel opens its Homer Research Laboratories in Bethlehem, Pa. At year's end, imports of foreign steel exceed exports.

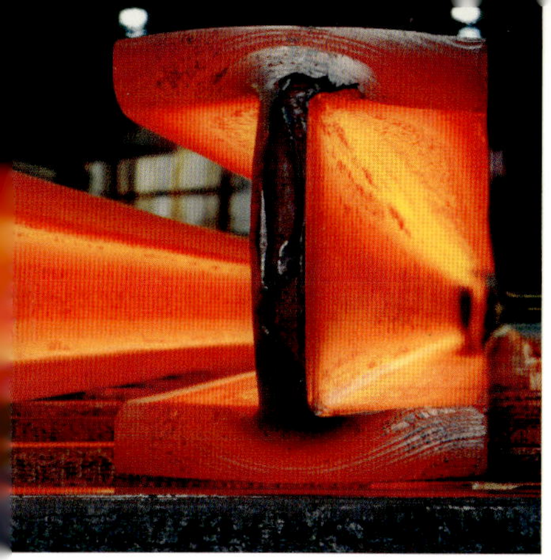

1964 Bethlehem Steel begins rolling steel at its new "greenfield" plant in Burns Harbor, Ind., along the southern shore of Lake Michigan.

1972 Martin Tower, the company's new headquarters building, was opened.

1977 Bethlehem Steel reports its first quarterly loss in 43 years, closes mills and lays off thousands of people.

1982 Losses mount, the Lackawanna plant is mostly closed, and Bethlehem Steel reports the largest corporate loss in American history.

1983 The number of people at work at Bethlehem Steel equals the number of retirees.

1988 Bethlehem Steel celebrates a wonderful year with net income of a record of $403 million.

1992 Bethlehem closes the Johnstown, Pa., plant that made bars, rods, and wire.

1993 New coating lines start up in Sparrows Point, Md., and Burns Harbor, Ind., solidifying Bethlehem's position as a leader in the production of light-flat-rolled steel for automotive, appliance, and metal-building applications.

1995 On a soulful morning on November 18, the fire at the last blast furnace at the home plant of Bethlehem Steel Corporation was extinguished, and metalmaking on the south shore of the Lehigh River ended after nearly 140 years.

1997 After being a component of the Dow Jones Industrial Index since 1918, Bethlehem Steel is eliminated from the index.

1998 Bethlehem Steel looked to the future when it bought Lukens Steel for $800 million, won after a bidding war with Allegheny Teledyne.

2000 With much fanfare, Bethlehem opens its last great investment – a new cold-rolling mill at Sparrows Point, Md.

2001 The fall of the stock market, the rising numbers of retirees and health care recipients, and the devastation to the American economy following the terrorist attacks of September 11th force Bethlehem Steel to seek bankruptcy protection on October 15.

2003 The assets, active, and idle, of Bethlehem Steel Corporation, are sold to International Steel Group.

Peter B. Treiber, Photographer

Peter Treiber is a formally trained photographer with an Associate Degree in Design from the State University of New York and a Bachelor of Fine Arts degree in Photography from the Art Center College of Design in California. After earning his undergraduate degree, he returned to New York City to work for three independent photographic studios. Peter was the Chief Photographer and Manager of Photographic Services at International Nickel Inc. in New York City before moving to Bethlehem, Pa., to work for Bethlehem Steel as an Advertising Photographer. Since 1983, Peter has owned a commercial photography studio in Bethlehem, Pa. (www.ptphoto.com)

Peter is a General Member and past Chapter President of the American Society of Media Photographers (ASMP) and a member of the National Association of Photoshop Professionals. During 24 years, Peter photographed operations at Bethlehem Steel and its clients' projects across America, resulting in an expansive collection of dramatic photographs.

Peter's commercial work has been published nationally and internationally. Exhibitions of his photography have been featured at Bethlehem City Hall under the auspices of the Bethlehem Fine Arts Commission, the Dakota Ridge Gallery, the Freedman Gallery at Albright College, the Hypno Gallery, the Iacocca Gallery at Lehigh University, Kutztown University, the Monsoon Gallery, the Payne Gallery at Moravian College, and by the ASMP Central Pennsylvania and the ASMP Philadelphia.

Elizabeth A. Kovach, Writer

Elizabeth, known as Bette, worked at Bethlehem Steel Corporation from early October 1976 until late May 2003. She served in corporate communications positions managing media relations, employee communications and publications, product marketing, executive speech writing, and issues management. Bette's last position with the company was Director of Corporate Communications where she served as the company spokesperson during times of extreme crisis. While with Bethlehem Steel, she also held leadership positions with industry-wide communications initiatives through the American Iron and Steel Institute (AISI), the industry's main trade association. With AISI, she chaired the communications committee and was Bethlehem Steel's lead representative for industry communications campaigns on trade and the enviroment.

Today, Bette manages her business, BK Communications, Inc. (www.bk-communications.com), which provides communications services to small-to-mid-size businesses and not-for-profit organizations. She has also consulted with large corporations on crisis and employee communications. Her clients span the legal, land development, financial services, and metals industries.

Bette holds a Bachelor of Arts degree in journalism from Moravian College and has completed extensive graduate work in business administration. She lectures on crisis communications, employee communications, and media relations. Bette continues to live and work in Bethlehem, Pa., where she remains very involved as a volunteer leader with not-for-profit organizations. She still marvels at the University of Pittsburgh's Cathedral of Learning, built by BSC in the shadow of U.S. Steel's headquarters, when she visits her daughter who attends the university.

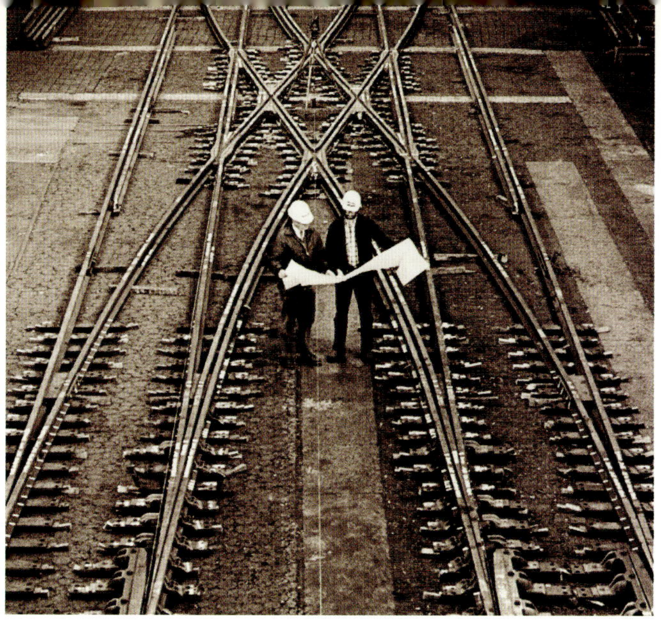

TECHNICAL NOTES

All but three photographs in this book were made with medium format, 2 1/4 x 2 1/4 cameras, using a variety of films and a digital back including black and white, color negative, and color transparency film. Bethlehem Steel only used negative films and made type "C" or black-and-white prints for reproduction. In preparing the images for this book, Peter Treiber photocopied a number of old "C" prints from his portfolio with his LightPhase camera back and cleaned them in Photoshop. He purchased digital scans of negatives to facilitate the publication of some earlier photos.

Most images are from scans Peter made of transparency film with a Minolta film scanner. His last few Bethlehem Steel projects were shot with a six mega-pixel LightPhase camera back and Hasselblad cameras. When cleaning up the photos in Photoshop, the differences in film type and quality quickly became evident. The old Verichrome Pan b&w and the LightPhase digital clearly stand out as the winners.

Fine Art Prints

Some of the photographs from this book have been exhibited as large, limited-edition, archival, fine art prints. Please consult Peter Treiber's website for the availability of prints for purchase or exhibit and for current gallery show information. (www.ptphoto.com)

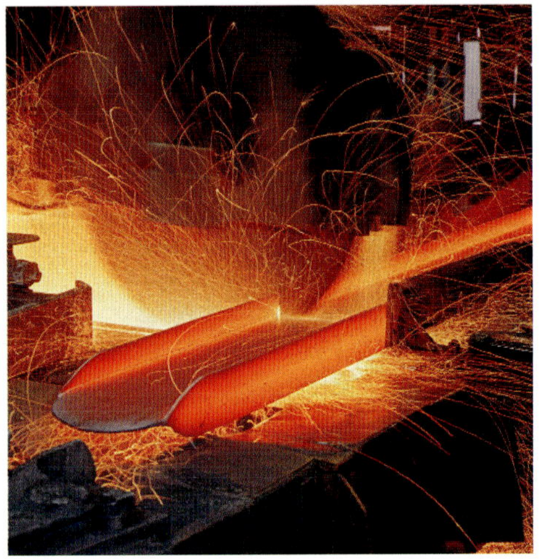

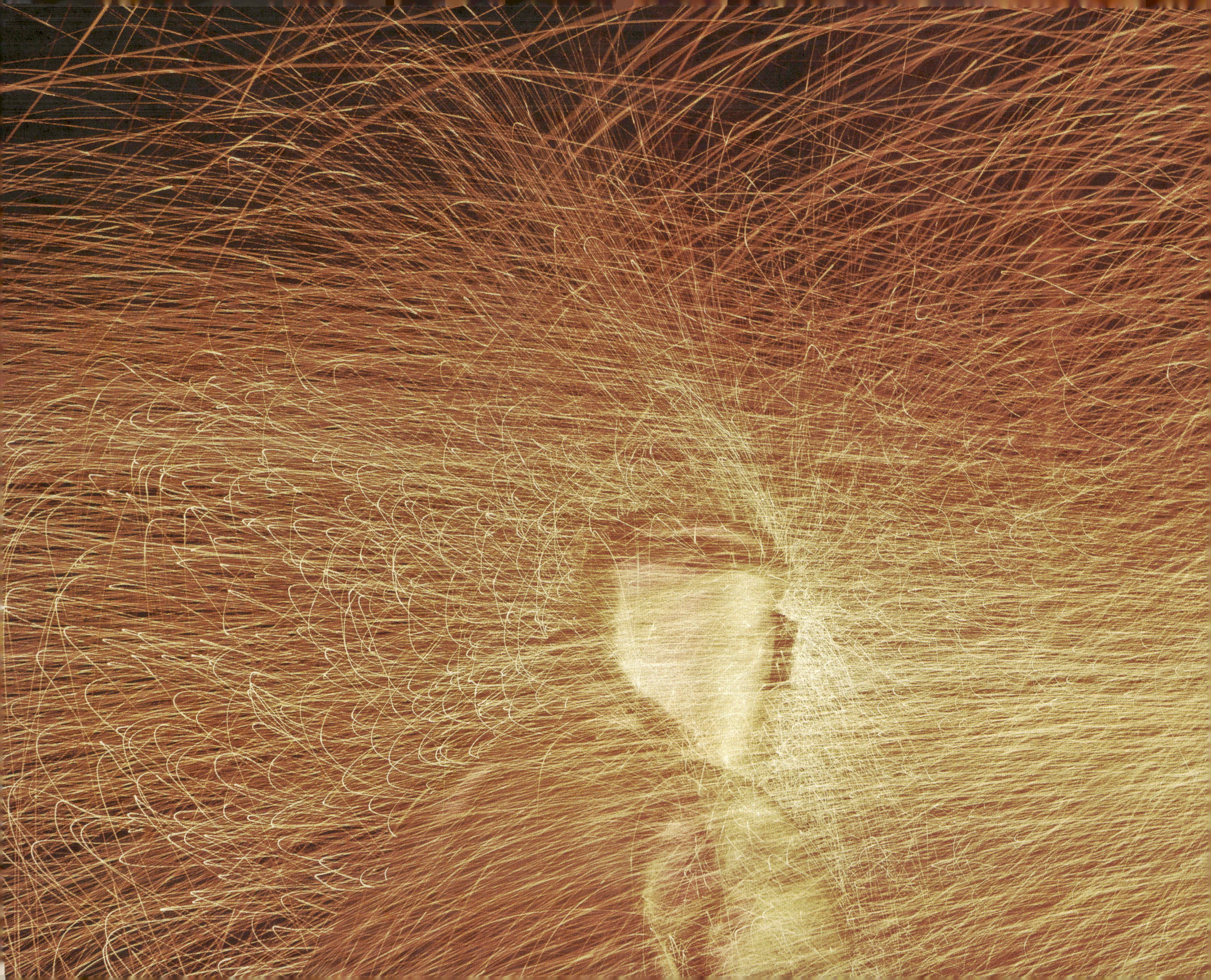

1 Bethlehem Steel made a wide variety of steel products for the defense industry, beginning with the Spanish American War and ending with the sale of the forging division in 1997 to West Homestead Engineering and Machinery Corporation of Homestead, Pa. In addition to this propeller shaft, the corporation made hull components, mechanical drive couplings, gears for gun tubes, and missile components. During World Wars I and II, Bethlehem forged gun barrels and turrets for battle ships as well as armored steel plates for ship hulls and other critical areas.

2 The forging division's laboratory services department provided a wide range of testing procedures including metallography and chemical analysis. The lab was an approved testing facility for the government and for the nuclear, chemical processing, aircraft, and aerospace industries. Bethlehem performed all testing and analytical work in accordance with required American Society for Testing and Materials specifications. Bethlehem's forging labs maintained a strong relationship with Bethlehem Steel's corporate Homer Research Laboratories, just minutes away up South Mountain in Bethlehem. Research employees were instrumental in the long-term success of forged products by developing new steel grades, introducing new metallurgical procedures, and guiding continued improvements in product quality.

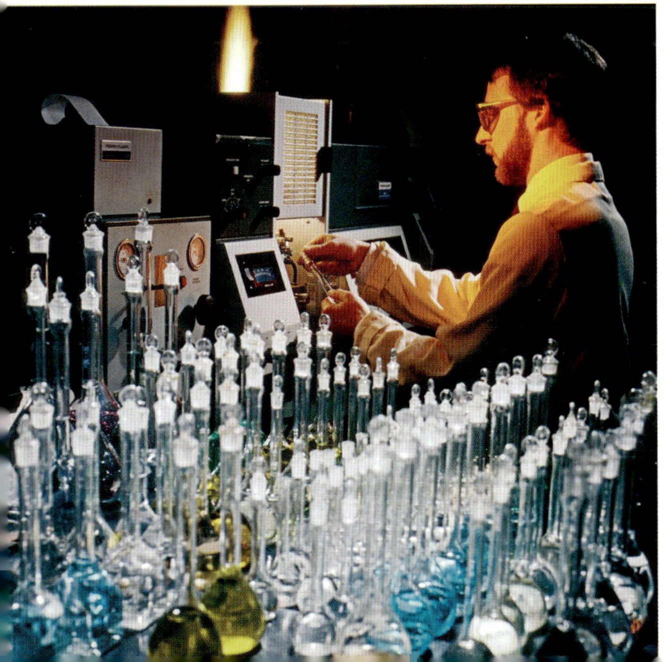

Metallurgical Lab

2

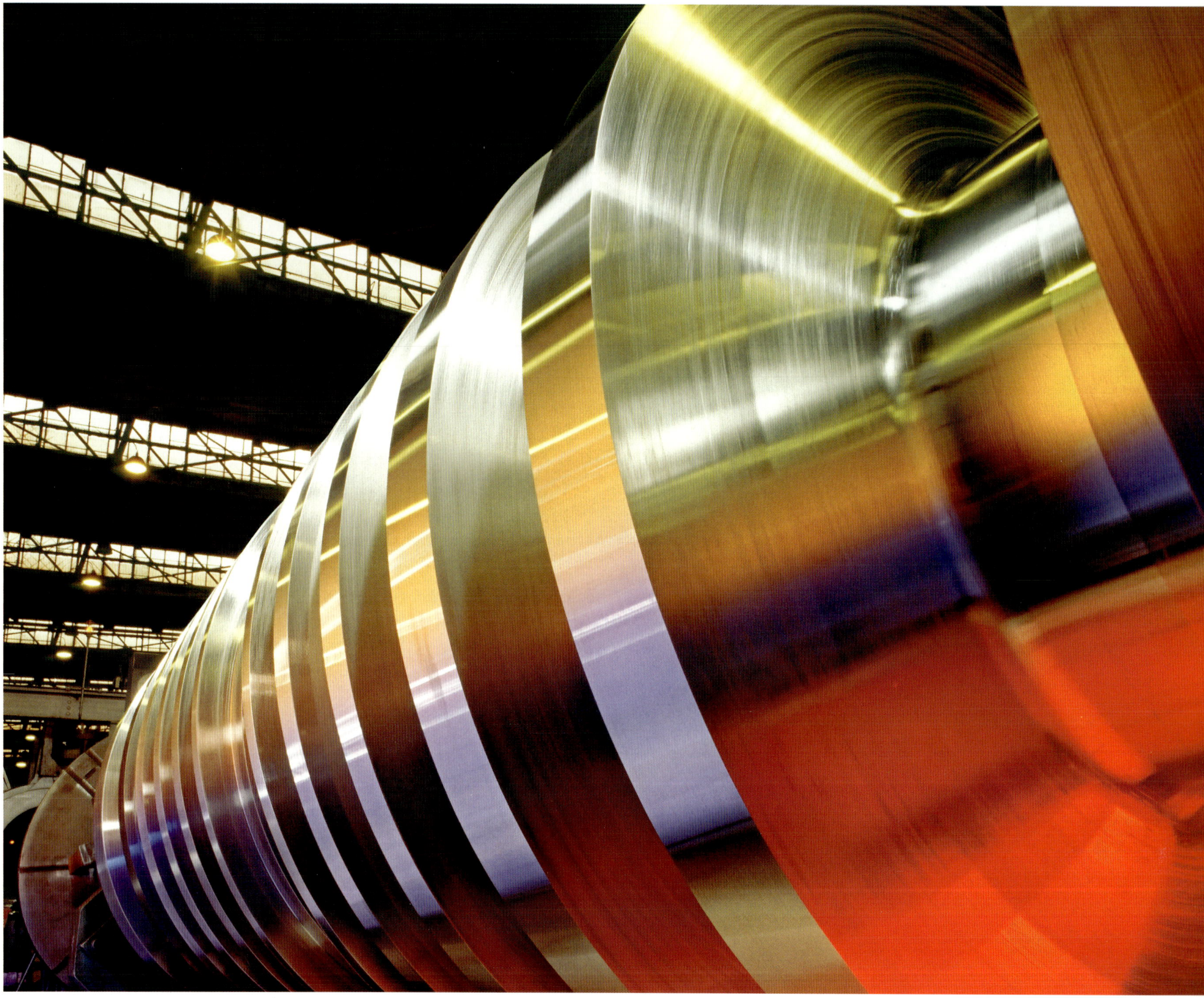

1

A turbine generator shaft in the rough

In the summer of 1999, the Bethlehem Fine Arts Commission asked me to participate in a large group show of Bethlehem Steel related art work as part of a city-wide "Steel Festival." This was good news since I had been wondering what I would ever do with the large numbers of BSC photographs in my files and my old, now unused, print portfolios. Judging by the turnout over the 10-day run and the accompanying press coverage, the festival was somewhat of a success. Some of us thought, "At last, someone cares about the history of Bethlehem Steel." At that time, plans for an industrial history museum and a retail and entertainment center on the unused plant site were languishing with no clear future. The company was still alive, but barely. Hope, in some quarters, remained that Bethlehem would once again prosper.

Therefore, contemplating and preserving BSC's history was somewhat premature; very little effort was being made to do anything about preserving material, such as photographs. While the struggle to survive was in progress, much of the living history, documents, and photographs were disappearing from the company's files and from the community.

This apparent lack of concern for the history of the company within the community was evident to me when no one purchased my hand-crafted fine art prints during the steel festival. So, for another seven years my BSC photos continued to be hidden away in my files. By 2006, any hope of seeing my images become part of the history of Bethlehem Steel had faded completely. I moved on and focused on serving my commercial accounts and on marketing my floral images through art and photography galleries.

In 2006 Dan Hugos, a photographer and owner of the Dakota Ridge Gallery in Jim Thorpe, Pa., decided to represent my floral photographs. Later conversations led to discussions of my work at Bethlehem Steel and Dan asked to see some of my steel images. I had made a few new 16" x 20" fine art prints and Dan was blown away. We soon planned a gallery show for 20 prints. I also showed the prints to Tom Werner, the head of the photography department at Parsons the New School for Design in New York City and a fine art consultant. Tom reinforced Dan's opinion that my photographs were both historically and artistically valuable. Editing for the Dakota Ridge show revealed a collection of well over 100 valuable photographs. A show of that size would not be economically feasible, but a book might be.

A book of my photos would not be complete without the text explaining the photographs and the history of Bethlehem Steel. Elizabeth "Bette" Kovach joined BSC in fall 1976, several months before I signed on, and persevered in the Public Affairs department to become the Director of Corporate Communications. We worked together as client and photographer on many enjoyable projects and had fun reminiscing about the good times. Bette, always the consummate professional, stayed on to the end in late May 2003. She knew BSC's history and is perfect for writing this book's text.

This book is not the complete history of Bethlehem Steel, but rather my vision of The Steel with Bette's description of the photos and historical background as she lived it. I am thankful that Bette has helped to make this book a reality and a truly enjoyable experience. We both hope that readers will gain a good understanding of what it was like inside Bethlehem Steel during the final quarter century.

Peter B. Treiber
Photographer

PREFACE

After I completed a one-month internship in Bethlehem Steel Corporation's public affairs department during my senior year at Moravian College in Bethlehem, Pa., my father warned me to never work for a big company. His disappointment came from working, ironically, for U.S. Steel's Federal Shipbuilding Division in New York Harbor near his, and later my, childhood home in West Orange, N.J., from 1938 until the yard closed in 1948. He loved his job and was devastated when the yard no longer had sustainable business after World War II.

But why would I listen to my father? Already enthralled by the majesty and might of the steel industry after one month, I joined Bethlehem Steel's public affairs department. The company offered a good salary, six weeks of paid vacation, and a full health and dental plan – all this to a young female in 1976.

My first two years at Bethlehem Steel were difficult as I tried to fit into an overwhelmingly engineering-oriented and male corporate culture. My innate competitiveness, however, helped me thrive, and I found my niche as a media representative and spokesperson in the home office, at the headquarters building named and known throughout the company as Martin Tower.

December 27, 1982 was the first hugely difficult test of my skills and fabric. On this day, Bethlehem Steel announced the closing of the nation's first fully integrated steel plant in Lackawanna, N.Y., signaling the beginning of a downward spiral of steel plant closings and sales of assets. Since 1979, I had worked with the Lackawanna, N.Y., plant, performing various media and employee communications functions and spending about one week a month at the plant. I knew the operation and its people well and was viewed by reporters and employees alike as Bethlehem Steel's lead media contact and spokesperson. Under intense pressure, I thrived on the challenge of delivering the news timely and objectively, and focused on the positive impact that closing a losing operation would have on the remaining facilities.

Over the years, I developed an affinity for "crisis communications," and, unfortunately, had many more occasions to deliver bad news. Like a slowly falling stack of dominoes, Bethlehem's assets, both large and small, were closed or sold during much of the corporation's last 20 years. A few good years held workers together and gave them eternal hope. The company spent billions of dollars in the 1980s and 1990s to stay competitive. Most people worked very hard to make Bethlehem successful. The downtrodden workforce was genuinely exhilarated when customers, including General Motors, lauded us for our success and when Bethlehem Steel opened new facilities.

The people at the plants had a gun to their heads every day. Make money, lower costs, increase productivity, and improve quality. Their jobs were hard, hot, and dangerous. But they were proud and truly loved seeing the fruits of their labors when steel was shipped bearing the familiar steel beam logo. During periods of turmoil, gallows humor throughout the company was evident in witty songs, poems, and cartoons that poked fun at the pain of trying to save a company that gave so much to so many people. A group of renegades even launched an ill-fated employee letter-writing campaign to convince the board of directors to select a chairman from outside the company for new, diverse leadership.

The 1990s had its ups and downs with a few good years and a few lousy ones. One of the most poignant days of my career came on November 18, 1995 when most of the Bethlehem, Pa., plant, the birthplace of the then-fading company, was closed after nearly 140 years of metalmaking on the south shore of the Lehigh River. It was a soulful, mournful, but great day to remember Bethlehem's legacy and to wish our coworkers well as they left with an enhancement to their pensions, deemed "a shutdown premium" by the United Steelworkers of America. Those who wanted to leave left and those who wanted to stay did so for a few years until the remaining operations at the Bethlehem plant were closed. That last day was like a block party, filled with tears, laughter, spouses, video cameras, and bar-be-cues, in the seemingly inhospitable environment of an old, gray steel plant. One blast furnace department fellow, slowing his car as he drove through the plant's main gate for the last time, said, on camera, "I bought two houses, two cars and put my two kids through college. Would I have liked Bethlehem Steel to last a little bit longer? Sure. But Bethlehem Steel was good to me. No regrets."

The mounting ratio of pensioners (74,000) to employees (12,000), the dot.com bust's impact on Wall Street, and the further devastation to the U.S. economy after the September 11th terrorist attacks brought Bethlehem to its knees, and, on October 15, 2001, the proud company that built America filed for Chapter 11 bankruptcy protection. Bethlehem wasn't the first steel company to file, but it was Bethlehem's filing that made the international news.

The board of directors explored a series of plans to keep the assets intact, and finally one was achievable. Late in the afternoon on Saturday, February 9, 2003, I sat in my office on the 17th floor of Martin Tower while the board of directors deliberated the sale of all Bethlehem Steel assets to newly formed International Steel Group.

On that same day, newspapers across the country featured a story of Bethlehem's announcement late the previous afternoon to terminate, effective March 31 of that year, health care benefits for more than 74,000 retirees and their eligible dependents. Over the years, I was the "name in the newspaper, the voice on the radio, and the face on the television" delivering bad news, and routinely received calls from all sorts of people to get more information about all sorts of things. But nothing, even the finality of the decisions being made that day, had prepared me for the call from an "older" lady from Wilmington, Del., being coached by her adult daughter whose voice I could hear in the background. Her soft voice cracked a bit when she said, "My daughter read in the paper today that I will no longer have medical insurance at the end of March. This cannot be! My late husband worked for 40 years at Bethlehem Steel. He worked at BETHLEHEM STEEL!" Neither she, nor I, nor thousands of others could believe that this great company would cease to exist.

In May 2003, still at work as the "last man standing" in public affairs, I tried to make sense of how a company just shy of 99 years could close. What happened to the momentum to live to celebrate 100 years, surely not an every-day occurrence in corporate America? I fully understood the reasons for the sale, the termination of health benefits, the termination of the defined benefit pension plan by the Pension Benefit Guaranty Corporation, and the need to sell facilities as on-going operations and not shut-down shells to retain the customer base. Sure, the remaining plants were operating, but the new owner didn't care one iota about tradition and legacy, and had, in the blink of an eye, removed all vestiges of the internationally recognizable I-beam symbol from every ongoing operation. The name and its symbol had been obliterated.

Ask one person what happened at Bethlehem Steel, and you will get one answer. Ask another, and you'll get another answer. Ask a third, a fourth, and a 100th – and the answers will all be different and equally debatable. In the end, we all knew the company was fragile. But we stayed nonetheless and tried to make a go of it. No one was chained to a coke-oven door, a blast-furnace sub car, a rolling-mill stand, or a desk. We all saw the cracks widening over the years and made our decisions to stay.

I am happy I did. I loved the steelmaking and finishing processes, the plants, and the people. It was a fabulous journey that provided tremendous challenges (a few of which I would have rather not encountered!), and I loved all things steel.

Writing *Inside Bethlehem Steel* has been a soulful and wonderful journey. The work brought back tons of memories, mostly happy, and rekindled my passion to protect Bethlehem, even in its grave. Working with Peter Treiber on this book of his outstanding photographs was a pleasure. His work filled Bethlehem Steel's brochures, website, advertisements, and product literature. Peter's attention to detail and ability to make the routine spectacular were recognized throughout the company, the industry, and the marketplace. He is creative, thoughtful, and respectful of his subjects. We worked well together then and now, and share a passion for a company long known for greatness.
Peter and I agree that Bethlehem Steel was a true example of American ingenuity that provided good-paying jobs to millions of people during nearly 99 years of production. Our intent is to celebrate the legacy and the successes of an American icon that was recognized throughout the world for its innovation and the employees' determination to be part of its legacy while living their American dream. Long live the memory of a company that produced the products that endure long after the mills are silent.

Elizabeth "Bette" Kovach

FOREWORD

In April of 1977, I went to work at Bethlehem Steel (BSC) as a staff advertising photographer in the mountain-top photographic services building. We were part of the in-house advertising agency, but were responsible for all of the company's photo needs. Bethlehem's six photographers worked for many departments all across the country and in a few foreign countries until the department's closure in 1983. After 1983, I continued photographing for BSC and its outside advertising agency as a freelancer until my last project in 2000.

The many perks and prestige of working at Bethlehem Steel were legendary, and my experience was similar to other salaried employees until the 1980s. On my first day in 1977 working at "The Steel," several of the other photographers took me out to lunch and told me that I was one of the luckiest photographers in the country. When I asked why, they told me that I had been selected from hundreds of applicants and that I now had a job for life with all of the cradle-to-grave benefits that went with it, including a company car and six weeks paid vacation a year. Having left a cushy job in New York City as the chief photographer for another mining and metal manufacturer, working at BSC was a rude awakening to the daily grind of photographing one dirty mill after another. This job meant constant travel by air and ground with countless hours of unpaid overtime. We had good equipment, but traveled light and without an assistant. For the photographers, the compensation and benefits were hard earned. The work was hot and dirty and provided Marty Forsher in New York City with a steady stream of Hasselblad camera repairs.

Although much of our work was used in advertising, it was definitely not a Madison Avenue type production. We were not shooting fashion models, but lots of filthy livestock barns, greasy steel mills, and low-ceiling coal mines. For many trips, we left on Sunday night and returned on Friday night or Saturday morning and traveled to a different city every day. Several times my wife met me at the airport with clean clothes and fresh film to start another week on the road. Sometimes the length of time spent on a project was much more important than the results.

Life was good in the 1970s, but the good life at "The Steel" for me only lasted about five years; our showplace photographic studio and labs were closed for good in 1983, sidelining a staff of 25 photographers and lab technicians. Built as part of the Homer Research Laboratory complex, the BSC photography department was a state-of-the-art facility. Eastman Kodak routinely brought in tours to show off our outstanding studio and labs.

In the early 1980s, Bethlehem Steel was reeling from new domestic and foreign competition, which was beating them on both quality and price. Bethlehem made a significant effort to upgrade and computerize the mills and to streamline the management and support staff. With that theme in mind, I made several attempts to market photo stories about the modernization of the American steel industry to various publications, but had no success. It seemed that no one was interested in the new steel making technology. As I continued to shoot steel projects for BSC's outside advertising agency, the game changed, allowing me to take a more creative approach using higher-end production techniques. As a freelancer, I switched to transparency film, did Polaroid testing, used lots of lighting, and always employed assistants. My last few projects were done with high-end digital systems.

ACNOWLEDGMENTS

Peter Treiber and Elizabeth Kovach would like to thank everyone who helped us to complete this wonderful journey resulting in *Inside Bethlehem Steel: The Final Quarter Century.* First, thanks to my wife Nancy Tilton Treiber for her tireless assistance and patience while I ignored my family responsibilities to pursue the seemingly endless work. Thanks also to Bette Kovach for her huge effort in writing the captions and to her husband Clay Morgan for his support of her career at Bethlehem Steel and her work on this book while she organized, wrote, and reorganized the text. Gratitude is also extended to Kathleen Forsythe who provided sterling design under pressure. Jenifer Gamber was a diligent and thorough editor. Bob Gallo gave us all attentive service at Hope Xinyuan Book Printing. Thanks also to the late Henry Von Spreckelsen, the former Manager of Corporate Communications, for so graciously permitting me to publish photos from my BSC portfolio over the years. Last, but not least, my thanks go to my many photo assistants who endured the dangerous and sometimes dirty work photographing at Bethlehem Steel with no complaints while providing superb service – Kelly Dillon, Jack Keefe, Doug Mason, David Moser, Michele Szoka, Jay Texter, and Steve Wolfe.

CONTENTS

DEDICATION

For nearly 100 years, more than one million employees of Bethlehem Steel Corporation created the steel that built, transported, and defended America. Particularly during the early years of the company, hundreds of people lost their lives working at Bethlehem's steelmaking, shipbuilding, coal mining, and manufacturing facilities. The years during World Wars I and II were glorious for the employees who supplied the steel to exacting specifications that helped the Allies win those wars, freeing millions of people from tyranny and launching a new age of prosperity in America. During the years of the company's long decline, bankruptcy, and eventual sale of its assets, tens of thousands of people lost their jobs, benefits, and part of their identity. This book is dedicated to the men and women who put their hearts and souls into their hard work while pursuing the American dream at Bethlehem Steel. Their work is a testimony to their exemplary contributions to this nation and the world. Both are better places because of them.

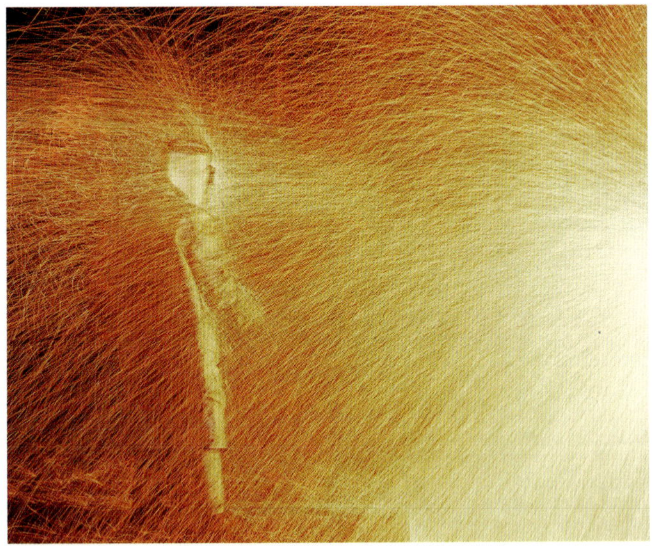